T0335563

OPTIMIZATION THEORY
A Concise Introduction

OPTIMIZATION THEORY

A Concise Introduction

Jiongmin Yong
University of Central Florida, USA

World Scientific

NEW JERSEY · LONDON · SINGAPORE · BEIJING · SHANGHAI · HONG KONG · TAIPEI · CHENNAI · TOKYO

Published by

World Scientific Publishing Co. Pte. Ltd.

5 Toh Tuck Link, Singapore 596224

USA office: 27 Warren Street, Suite 401-402, Hackensack, NJ 07601

UK office: 57 Shelton Street, Covent Garden, London WC2H 9HE

British Library Cataloguing-in-Publication Data
A catalogue record for this book is available from the British Library.

OPTIMIZATION THEORY
A Concise Introduction

Copyright © 2018 by World Scientific Publishing Co. Pte. Ltd.

All rights reserved. This book, or parts thereof, may not be reproduced in any form or by any means, electronic or mechanical, including photocopying, recording or any information storage and retrieval system now known or to be invented, without written permission from the publisher.

For photocopying of material in this volume, please pay a copying fee through the Copyright Clearance Center, Inc., 222 Rosewood Drive, Danvers, MA 01923, USA. In this case permission to photocopy is not required from the publisher.

ISBN 978-981-3237-64-3

For any available supplementary material, please visit
http://www.worldscientific.com/worldscibooks/10.1142/10923#t=suppl

Printed in Singapore

To

Meifen and Our Kids

Preface

In daily life, people frequently face problems on decision-making. Among all possible decisions, it is desired to select the best one. Is there the best (or optimal) decision? If yes, how can one find it (efficiently)? Optimization theory is to answer such kinds of questions to some extent. Mathematically, the problem can be formulated as follows. For a given objective function defined on the set of all decision variables, find a decision such that the objective is optimized (minimized or maximized). As a matter of fact, minimization and maximization problems were one of the main motivations for the invention of calculus back in the 17th century. Subject-wise, optimization theory can actually be regarded as a natural continuation of (advanced) calculus.

Needless to say, having some basic knowledge of optimization theory will probably enable people to wisely handle various encountered decision-making problems. On the other hand, except those researchers in the area of optimization theory, most people might only be interested in knowing some basics of optimization theory. Therefore, it should be ideal to have a one-semester course that could cover some basic topics in optimization, including linear, convex, and nonlinear optimization. This is my motivation of writing this book to meet such a need. In the past few years, I have taught such a course at University of Central Florida several times, in two different levels, for the upper level undergraduate students and first or second year graduate students. This book is based on my lecture notes of the course that I had taught.

This book is intended to present a brief introduction to optimization theory. After recalling some basic knowledge of advanced calculus and linear algebra, we start the presentation of general optimization problems, followed by abstract Weierstrass type existence theorems. Then we introduce the Lagrange multiplier method for optimization problems with equality constraints in high dimensions (two- and three-dimensional versions of which are introduced in the standard calculus courses). After that, Karush–Kuhn–Tucker method for problems with equality and inequality constraints is presented. In the case that no regularity condition for the constraints is assumed, we present Fritz John's necessary conditions for optimal solutions, with the proof by using Ekeland's variational principle (such a method has been used in establishing Pontryagin's maximum principle in optimal control theory, and to my best knowledge, this method has not been used in the context similar to this book). Next, we present optimization problems under convex/quasi-convex constraints. Some basic important results in convex analysis are presented, such as Carathéodory theorem of convex hull representation, separation theorem of convex sets, Lipschitz continuity of convex functions, sufficiency of the Karush–Kuhn–Tucker conditions for the problems under convex/quasi-convex constraints, Lagrange duality, and so on. Finally, we present a self-contained theory of linear programming. Characterization of extreme points for the feasible set is presented, weak and strong versions of fundamental theorem of linear programming are established. Simplex method, including phase I and phase II, as well as sensitivity analysis, and duality theory are presented.

The book is designed to be suitable for a one-semester course either for upper level undergraduate students, or first/second year graduate students. For the former, the instructor is recommended to concentrate on the computational/calculative aspect of the material, and skip most proofs of theorems. For the latter, the instructor is suggested to spend less time on the computational aspect, very briefly mention the results from the first chapter, and concentrate on most, if not all, proofs of the theorems.

Orlando, FL

Jiongmin Yong
January 2018

Contents

Chapter 1

Mathematical Preparations

In this chapter, we will briefly present some basic results of advanced calculus (analysis) and linear algebra, which will be used in the later chapters. We will mainly concentrate on the definitions and statements of relevant results, with short proofs. Lengthy proofs will be omitted, and they can be found in standard textbooks of analysis and linear algebra. Also, we do not try to be exhaustive. There are two types of readers that can skip this chapter: Those who have solid knowledge of analysis and linear algebra, and those who are only interested in computational aspects of optimization theory.

1 Basics in $\mathbb{R}^n$

Let $\mathbb{R}$ be the sets of all real numbers. For any $n \geqslant 1$, we define

$$\mathbb{R}^n = \left\{ \begin{pmatrix} x_1 \\ \vdots \\ x_n \end{pmatrix} \mid x_i \in \mathbb{R}, \quad i = 1, 2, \cdots, n \right\},$$

and identify $\mathbb{R}^1 = \mathbb{R}$. Elements in $\mathbb{R}^n$ are called *vectors* and briefly denoted by x, y, z, etc., i.e.,

$$x = \begin{pmatrix} x_1 \\ \vdots \\ x_n \end{pmatrix}, \qquad y = \begin{pmatrix} y_1 \\ \vdots \\ y_n \end{pmatrix}, \qquad z = \begin{pmatrix} z_1 \\ \vdots \\ z_n \end{pmatrix}.$$

For notational convenience and presentation consistency, we will let all vectors in $\mathbb{R}^n$ be *column vectors*, and denote

$$(x_1, \cdots, x_n) = \begin{pmatrix} x_1 \\ \vdots \\ x_n \end{pmatrix}^\top,$$

and call it a *row vector*, with the superscript $\top$ being called the *transpose*.

Now, let us introduce some *binary relations* between two vectors. For any $x \equiv (x_1, \cdots, x_n)^\top, y \equiv (y_1, \cdots, y_n)^\top \in \mathbb{R}^n$, we write

$$x = y, \quad \text{if} \quad x_i = y_i, \quad i = 1, 2, \cdots, n,$$
$$x \geqslant y, \quad \text{if} \quad x_i \geqslant y_i, \quad i = 1, 2, \cdots, n,$$
$$x > y, \quad \text{if} \quad x_i > y_i, \quad i = 1, 2, \cdots, n.$$

Note that when $n = 1$, for any $x, y \in \mathbb{R}$, one must have

$$\text{either} \quad x \geqslant y, \quad \text{or} \quad y \geqslant x. \tag{1.1}$$

We refer to the relation "$\geqslant$" an *order* in $\mathbb{R}$. Further, we can refine the above as follows: For any $x, y \in \mathbb{R}$, one and only one of the following must be true:

$$x = y, \quad x > y, \quad y > x. \tag{1.2}$$

However, if $n \geqslant 2$, there are $x, y \in \mathbb{R}^n$ for which none of (1.2) holds. For example, for $x = (1, 0)^\top$ and $y = (0, 1)^\top$, none of the relations in (1.1) holds. Hence, for the case $n \geqslant 2$, we usually refer to the relation "$\geqslant$" as a *partial order*, since it only gives relations to some pairs of vectors. Next, we define

$$\mathbb{R}^n_+ = \{x \in \mathbb{R}^n \mid x \geqslant 0\}, \qquad \mathbb{R}^n_{++} = \{x \in \mathbb{R}^n \mid x > 0\},$$

which are called *non-negative* and *strictly positive orthants* of $\mathbb{R}^n$, respectively. It is clear that

$$x = y, \qquad \Longleftrightarrow \qquad x \geqslant y \quad \text{and} \quad y \geqslant x.$$

Here, " $\Longleftrightarrow$ " means "if and only if". Also, we define

$$x \leqslant y, \quad \text{if} \quad y \geqslant x,$$
$$x < y, \quad \text{if} \quad y > x.$$

Let us look at some basics on the *sets* in $\mathbb{R}^n$. If A is a set in $\mathbb{R}^n$ consisting of vectors $x \in \mathbb{R}^n$ satisfying certain property, denoted by (P), then we write

$$A = \{x \in \mathbb{R}^n \mid x \text{ satisfies (P)}\}.$$

For example,

$$A = \{x \in \mathbb{R}^2 \mid x \geqslant 0\}. \tag{1.3}$$

If A is a set in $\mathbb{R}^n$, and a vector x is an element of A, we denote $x \in A$, and say "x belongs to A", or "x is in A", otherwise, we denote $x \notin A$, and say "x is not in A". For example, if A is defined by (1.3), then

$$\begin{pmatrix} 1 \\ 1 \end{pmatrix} \in A, \qquad \begin{pmatrix} 1 \\ -1 \end{pmatrix} \notin A.$$

If A and B are two sets in $\mathbb{R}^n$ such that any elements in A is in B, we call A a *subset* of B, denoted by $A \subseteq B$. If $A \subseteq B$ and there is at least one element which is in B but not in A, then A is called a *proper subset* of B, denoted by $A \subset B$. Two sets are *equal* if $A \subseteq B$ and $B \subseteq A$. The set that contains no element is called the *empty set*, denoted by $\varnothing$. Let A_λ be a family of sets in $\mathbb{R}^n$ parameterized by $\lambda \in \Lambda$, where Λ is an index set which could be a finite set or an infinite set. We define

$$\bigcup_{\lambda \in \Lambda} A_\lambda \triangleq \Big\{ x \in \mathbb{R}^n \mid x \in A_\lambda, \text{ for some } \lambda \in \Lambda \Big\},$$

$$\bigcap_{\lambda \in \Lambda} A_\lambda \triangleq \Big\{ x \in \mathbb{R}^n \mid x \in A_\lambda, \text{ for all } \lambda \in \Lambda \Big\},$$

and they are called the *union* and *intersection* of A_λ ($\lambda \in \Lambda$), respectively. Also, for any $A, B \subseteq \mathbb{R}^n$, we define

$$A \setminus B \triangleq \{ x \in A \mid x \notin B \},$$

which is called "A minus B". As a special case, we denote $B^c = \mathbb{R}^n \setminus B$, which is called the *complementary* of B. It is clear that

$$A \subseteq B \quad \Longleftrightarrow \quad B^c \subseteq A^c.$$

We have the following result.

Proposition 1.1. (de Morgan's Law) *Let* $\{A_\lambda, \lambda \in \Lambda\}$ *be a family of sets in* $\mathbb{R}^n$. *Then*

$$\Big(\bigcup_{\lambda \in \Lambda} A_\lambda \Big)^c = \bigcap_{\lambda \in \Lambda} A_\lambda^c, \quad \Big(\bigcap_{\lambda \in \Lambda} A_\lambda \Big)^c = \bigcup_{\lambda \in \Lambda} A_\lambda^c. \tag{1.4}$$

Proof. Look at the following:

$$x \in \Big(\bigcup_{\lambda \in \Lambda} A_\lambda \Big)^c \quad \Longleftrightarrow \quad x \notin \bigcup_{\lambda \in \Lambda} A_\lambda \quad \Longleftrightarrow \quad x \notin A_\lambda, \ \forall \lambda \in \Lambda$$

$$\Longleftrightarrow \quad x \in A_\lambda^c, \ \forall \lambda \in \Lambda \quad \Longleftrightarrow \quad x \in \bigcap_{\lambda \in \Lambda} A_\lambda^c,$$

proving the first relation in (1.4). Similarly, we can prove the second one, which is left to the readers. $\qquad\blacksquare$

We now introduce two *operations*:

- *Addition*: For any $x = (x_1, \cdots, x_n)^\top, y = (y_1, \cdots, y_n)^\top \in \mathbb{R}^n$, define

$$x + y = (x_1 + y_1, \cdots, x_n + y_n)^\top. \tag{1.5}$$

● *Scalar multiplication*: For any $x = (x_1, \cdots, x_n)^\top \in \mathbb{R}^n$ and $\lambda \in \mathbb{R}$, define

$$\lambda x = (\lambda x_1, \cdots, \lambda x_n)^\top. \tag{1.6}$$

The above are called *linear operations* (or *algebraic operations*) on $\mathbb{R}^n$. Inspired by the above, we introduce some abstract notions which will give us a better view on more general problems.

Definition 1.2. A non-empty set X is called a *linear space* (over $\mathbb{R}$) if on X, two operations are defined: one is called *addition*:

$$(x, y) \mapsto x + y \in X, \qquad \forall x, y \in X,$$

and the other is called *scalar multiplication*:

$$(\lambda, x) \mapsto \lambda x \in X, \qquad \forall \lambda \in \mathbb{R}, \ x \in X.$$

These operations satisfy the following *eight* axioms:

(i) For any $x, y \in X$, $x + y = y + x$;

(ii) For any $x, y, z \in X$, $(x + y) + z = x + (y + z)$;

(iii) There exists a unique element called zero in X, denoted by 0, such that $x + 0 = 0 + x = x$, for all $x \in X$;

(iv) For any $x \in X$, there exists a unique $y \in X$ such that $x + y = y + x = 0$. Such a y is denoted by $-x$;

(v) For any $x, y \in X$ and $\lambda \in \mathbb{R}$, $\lambda(x + y) = \lambda x + \lambda y$;

(vi) For any $\lambda, \mu \in \mathbb{R}$ and $x \in X$, $(\lambda + \mu)x = \lambda x + \mu x$;

(vii) For any $x \in X$, $1x = x$;

(viii) For any $\lambda, \mu \in \mathbb{R}$ and $x \in X$, $(\lambda\mu)x = \lambda(\mu x)$.

A linear space is also called a *vector space*. Any element in a linear space is called a *vector*.

As an important example, we have that $\mathbb{R}^n$ is a linear space under the addition and scalar multiplication defined by (1.5) and (1.6)[1].

If $X = \{0\}$, we call it the *zero linear space* whose structure is trivial. If a linear space $X \neq \{0\}$, we call it a *non-zero linear space*. To study the

---

[1]There are many other linear spaces, besides $\mathbb{R}^n$, for example, the set of all $(n \times m)$ matrices, the set of all polynomials, the set of all continuous functions, to mention a few.

structure of a non-zero linear space X (which could be $\mathbb{R}^n$), we introduce the following important notions.

Definition 1.3. Let X be a non-zero linear space over $\mathbb{R}$.

(i) A set $\{\xi_1, \xi_2, \cdots, \xi_k\} \subset X$ of finitely many vectors is said to be *linearly dependent* if there are constants $\lambda_1, \cdots, \lambda_k \in \mathbb{R}$, at least one of them is non-zero, such that

$$\lambda_1 \xi_1 + \lambda_2 \xi_2 + \cdots + \lambda_k \xi_k \equiv \sum_{i=1}^{k} \lambda_i \xi_i = 0. \tag{1.7}$$

Otherwise, these vectors are said to be *linearly independent*. The left hand side of (1.7) is called a *linear combination* of $\xi_1, \cdots, \xi_k$.

(ii) A set $\{\xi_1, \cdots, \xi_k\} \subset X$ is called a *basis* of X if they are linearly independent and for any $x \in X$, there are constants $\lambda_1, \cdots, \lambda_k \in \mathbb{R}$ such that

$$x = \sum_{i=1}^{k} \lambda_i \xi_i. \tag{1.8}$$

(iii) A subset $X_0 \subseteq X$ is called a (linear) *subspace* of X if

$$\lambda x + \mu y \in X_0, \qquad \forall x, y \in X_0, \ \lambda, \mu \in \mathbb{R}.$$

For any linear space X, there are two trivial subspaces: X and $\{0\}$. Other subspaces are called *proper* subspaces. The following result is fundamentally important.

Proposition 1.4. *Let X be a non-zero linear space over $\mathbb{R}$.*

(i) *Let $\xi_1, \cdots, \xi_k \in X$. Then the span of $\{\xi_1, \cdots, \xi_k\}$ defined by*

$$\text{span}\,\{\xi_1, \cdots, \xi_k\} = \Big\{ \sum_{i=1}^{k} \lambda_i \xi_i \ \Big| \ \lambda_i \in \mathbb{R}, \ 1 \leqslant i \leqslant k \Big\},$$

is a linear subspace of X.

(ii) *If $X_0 \neq \{0\}$ is a subspace of X, then there is a set $\{\xi_1, \cdots, \xi_k\} \subset X_0$ of linearly independent vectors such that*

$$X_0 = \text{span}\,\{\xi_1, \cdots, \xi_k\}. \tag{1.9}$$

In this case, $\{\xi_1, \cdots, \xi_k\}$ must be a basis of X_0. Further, if $\{\eta_1, \cdots, \eta_\ell\}$ is another basis of X_0, then it must be true that $\ell = k$.

(iii) *If $\{\xi_1, \cdots, \xi_k\}$ is a basis of X, then any vector $x \in X$ admits a unique representation of form (1.8).*

Because of (ii) in Proposition 1.4, we call k the *dimension* of X, denoted by $\dim X = k$. This is the maximum possible number of linearly independent vectors in the space X. If we let $e_i \in \mathbb{R}^n$ be the vector with the i-th component being 1 and all other components being zero, then $\{e_1, \cdots, e_n\}$ is a basis of $\mathbb{R}^n$, which is called the *canonical basis* of $\mathbb{R}^n$. From this, we see that $\dim \mathbb{R}^n = n$.

Note that to find a basis for a non-zero linear space X, one can start from any non-zero vector, call it ξ_1. If any other vector in X is a multiple of ξ_1, we are done and $\{\xi_1\}$ is a basis of X. Otherwise, one can find a vector $\xi_2 \in X$ which is not a multiple of ξ_1. Then ξ_1 and ξ_2 have to be linearly independent. If any vector in X is a linear combination of ξ_1 and ξ_2, we are done and $\{\xi_1, \xi_2\}$ is a basis of X. Otherwise, we can continue the above procedure, and eventually, we will find a basis of X. If the above procedure stops at some finite step, i.e., $\dim X = k$ for some k, we say that X is a *finite-dimensional space*. Otherwise, X is called an *infinite-dimensional space*[2]. Note that for a given linear space, the basis is not unique.

From the above discussion, we see that thanks to the linear operations of $\mathbb{R}^n$ and the notion of basis, the structure of any subspace X_0 (including $\mathbb{R}^n$ itself) of $\mathbb{R}^n$ becomes very simple: As long as a basis $\{\xi_1, \cdots, \xi_k\}$ of X_0 is found, X_0 can be represented by (1.9). We refer to the above description as the *linear structure* (or *algebraic structure*) of $\mathbb{R}^n$.

Next, on $\mathbb{R}^n$, we introduce two more notions:

- *Dot product*: For any $x = (x_1, \cdots, x_n)^\top, y = (y_1, \cdots, y_n)^\top \in \mathbb{R}^n$, let

$$x \cdot y \equiv x_1 y_1 + x_2 y_2 + \cdots + x_n y_n \equiv \sum_{i=1}^{n} x_i y_i. \tag{1.10}$$

- *Euclidean norm*: For any $x = (x_1, \cdots, x_n)^\top \in \mathbb{R}^n$, define

$$\|x\|_2 = \Big(\sum_{i=1}^{n} |x_i|^2 \Big)^{\frac{1}{2}}. \tag{1.11}$$

More generally, we introduce the following abstract notions.

Definition 1.5. Let X be a linear space over $\mathbb{R}$.

(i) A binary operation, denoted by $(x, y) \mapsto x \cdot y$ from $X \times X$ to $\mathbb{R}$, is called an *inner product* if the following are satisfied:

---

[2]For example, the set of all continuous functions defined on, say, $[0, 1]$, is an infinite-dimensional linear space.

(a) *positivity:* For any $x \in X$, $x \cdot x \geqslant 0$, and $x \cdot x = 0$ if and only if $x = 0$;

(b) *symmetry:* For any $x, y \in X$, $x \cdot y = y \cdot x$;

(c) *linearity:* For any $x, y, z \in X$, and $\lambda, \mu \in \mathbb{R}$, $(\lambda x + \mu y) \cdot z = \lambda(x \cdot z) + \mu(y \cdot z)$.

When a linear space X admits an inner product, it is called an *inner product space*.

(ii) A map $x \mapsto \|x\|$ from X to $\mathbb{R}$ is called a *norm* if the following are satisfied:

(a) *positivity:* For any $x \in X$, $\|x\| \geqslant 0$ and $\|x\| = 0$ if and only if $x = 0$;

(b) *positive homogeneity:* For any $\lambda \in \mathbb{R}$ and $x \in X$, $\|\lambda x\| = |\lambda| \, \|x\|$;

(c) *triangle inequality:* For and $x, y \in X$ and $\|x + y\| \leqslant \|x\| + \|y\|$.

When a linear space X admits a norm, it is called a *normed linear space*.

The following proposition reveals some relations between inner product and norm.

Proposition 1.6. *Let X be an inner product space.*

(i) **(Cauchy–Schwarz inequality)** *For any $x, y \in X$, it holds*

$$|x \cdot y|^2 \leqslant (x \cdot x)(y \cdot y). \tag{1.12}$$

The equality holds if and only if x and y are linearly dependent.

(ii) *The map $x \mapsto \|x\|$ defined by the following:*

$$\|x\| = \sqrt{x \cdot x}, \qquad \forall x \in X$$

is a norm on X, under which X is a normed linear space. In this case $\| \cdot \|$ is called the norm induced by the inner product.

Note that a norm is not necessarily induced by an inner product.

Proof. (i) First of all, if $y = 0$, (1.12) is trivially true. Now, let $y \neq 0$. For any $\lambda \in \mathbb{R}$, we have

$$0 \leqslant (x + \lambda y) \cdot (x + \lambda y) = (x \cdot x) + 2(x \cdot y)\lambda + (y \cdot y)\lambda^2 \equiv f(\lambda).$$

In particular,

$$0 \leqslant f\left(-\frac{x \cdot y}{y \cdot y}\right) = (x \cdot x) - \frac{(x \cdot y)^2}{(y \cdot y)}.$$

Hence, (1.12) follows.

The rest of the proof is left to the readers.   □

From the above, we see that the Euclidean norm $\|\cdot\|_2$ defined by (1.11) is a norm on $\mathbb{R}^n$, which is induced by the dot product defined by (1.10), and $\mathbb{R}^n$ is an inner product space under the dot product and a normed linear space under $\|\cdot\|_2$. In what follows, we call $\mathbb{R}^n$ a *Euclidean space* if the linear operations are defined by (1.5)–(1.6), and the inner product and the norm are defined by (1.10)–(1.11).

Having the norm $\|\cdot\|$ in a linear space X, we can define the *distance* between two vectors $x, y \in X$ by $\|x - y\|$. We call $\|x - y\|_2$ the *Euclidean distance* between $x, y \in \mathbb{R}^n$, which is a natural generalization of the usual distances on the line, on the plane, and in the 3-dimensional space. Clearly, $\|x\|_2$ is nothing but the distance of the vector x to the zero vector $0 = (0, \cdots, 0)^\top$, which can also be called the *length* of the vector x. Note that when $n = 1$, $\|x\|_2$ is nothing but the *absolute value* of $x \in \mathbb{R}$:

$$|x| = \begin{cases} x, & x \geqslant 0, \\ -x, & x < 0. \end{cases}$$

A vector x in a normed linear space X is called a *unit vector* if $\|x\| = 1$. Clearly, a vector x in a normed linear space is non-zero if and only if $\|x\| > 0$.

From the Cauchy–Schwarz inequality, we can define the *angle* $\theta \in [0, \pi]$ between two non-zero vectors x and y in an inner product space X through the following:

$$\cos \theta = \frac{x \cdot y}{\|x\| \, \|y\|}.$$

Then two vectors $x, y \in X$ are said to be *orthogonal*, denoted by $x \perp y$, if $x \cdot y = 0$ (which is equivalent to $\cos \theta = 0$ with θ being the angle between these two vectors). For any finite-dimensional inner product space X, if $\{\xi_1, \cdots, \xi_k\}$ is a basis of X such that

$$\xi_i \cdot \xi_j = \begin{cases} 1, & i = j, \\ 0, & i \neq j, \end{cases}$$

which means that $\{\xi_1, \cdots, \xi_k\}$ is a set of mutually orthogonal unit vectors, we call $\{\xi_1, \cdots, \xi_k\}$ an *orthonormal basis* of X. It is clear that $\{e_1, \cdots, e_n\}$ is an orthonormal basis of $\mathbb{R}^n$ under the dot product. Note that for a given inner product space, the orthonormal basis is not unique. The following result provides some facts about the orthogonality.

Proposition 1.7. *Let X be an inner product space.*

(i) *Suppose G is a non-empty subset of X. Then the annihilator $G^\perp$ of G defined by*

$$G^\perp \triangleq \{y \in X \mid y \cdot x = 0, \ \forall x \in G\}$$

is a subspace of X, and

$$(G^\perp)^\perp = \text{span}\,(G) = \left\{ \sum_{i \geqslant 1} \lambda_i \xi_i \mid \xi_i \in G, \ \lambda_i \in \mathbb{R}, \ i \geqslant 1 \right\} \supseteq G. \quad (1.13)$$

(ii) *If $\varnothing \neq G_1 \subseteq G_2 \subseteq X$, then*

$$G_1^\perp \supseteq G_2^\perp. \quad (1.14)$$

(iii) *If G is a subspace of X, then*

$$(G^\perp)^\perp = G, \quad (1.15)$$

and

$$X = G + G^\perp \equiv \{x + y \mid x \in G, \ y \in G^\perp\}, \quad G \cap G^\perp = \{0\}. \quad (1.16)$$

The proof is left to the readers. Note that by definition, one has

$$x \perp y, \qquad \forall x \in G, \ y \in G^\perp.$$

Thus, we could simply write

$$G^\perp \perp G. \quad (1.17)$$

When the two properties in (1.16) hold, we write

$$X = G \oplus G^\perp,$$

and call it an *orthogonal decomposition* of X. Hence, in the case G is a subspace of X, we call $G^\perp$ the *orthogonal complementary* of G.

To be specific, in the rest of this subsection, we will only consider Euclidean space $\mathbb{R}^n$. Now, we introduce the following definition.

Definition 1.8. (i) The *open ball* $B_r(x)$ and the *closed ball* $\bar{B}_r(x)$ centered at x with radius r are defined by the following:

$$B_r(x) = \left\{y \in \mathbb{R}^n \mid \|y - x\|_2 < r\right\}, \quad \bar{B}_r(x) = \left\{y \in \mathbb{R}^n \mid \|y - x\|_2 \leqslant r\right\};$$

(ii) A set $G \subseteq \mathbb{R}^n$ is called an *open set* if for any $x \in G$, there exists a $\delta > 0$ such that

$$B_\delta(x) \subseteq G.$$

(ii) A set $G \subseteq \mathbb{R}^n$ is called a *closed set* if $G^c \equiv \mathbb{R}^n \setminus G$ is open.

(iii) The *interior* G° of a set $G \subseteq \mathbb{R}^n$ is the largest open set that is contained in G; the *closure* $\bar{G}$ of a set $G \subseteq \mathbb{R}^n$ is the smallest closed set that contains G; the *boundary* ∂G of G is defined as follows:

$$\partial G = \bar{G} \bigcap \overline{G^c}.$$

(iv) G is called a *bounded set* if there exists a constant K such that $G \subseteq B_K(0)$; G is called a *compact set* if it is bounded and closed[3].

(v) G is said to be *connected* if there are no non-empty subsets $G_1, G_2 \subset G$ such that

$$G = G_1 \cup G_2, \qquad \bar{G}_1 \cap G_2 = \varnothing, \qquad G_1 \cap \bar{G}_2 = \varnothing.$$

(vi) G is called a *domain* if it is open and connected.

Note that the dot product leads to the norm, the distance, and the notion of open balls, then leads to the open sets in $\mathbb{R}^n$ (and hence closed sets, bounded sets, compact sets, etc.). Anything related to open sets, distance, and so on, is referred to as a *topological property*. Therefore, the dot product, the norm, or the distance determines the *topological structure* of $\mathbb{R}^n$. Now, we consider sequences in $\mathbb{R}^n$. The most important notion for sequences is the following:

Definition 1.9. A *sequence* $\{x_k\}_{k \geqslant 1} \subset \mathbb{R}^n$ is *convergent* to $\bar{x} \in \mathbb{R}^n$, denoted by $x_k \to \bar{x}$ or $\lim_{k \to \infty} x_k = \bar{x}$, if for any $\varepsilon > 0$, there exists a $K \geqslant 1$ (depending on $\varepsilon > 0$) such that

$$\|x_k - \bar{x}\|_2 < \varepsilon, \qquad \forall k \geqslant K.$$

In this case, $\bar{x}$ is called the *limit* of the sequence $\{x_k\}_{k \geqslant 1}$.

The following proposition collects relevant results concerning the sequences in $\mathbb{R}^n$, whose proof is left to the readers.

Proposition 1.10. *Let* $\{x_k\}_{k \geqslant 1}$ *be a sequence in* $\mathbb{R}^n$. *Then the following statements are equivalent:*

---

[3]This definition only works for finite-dimensional normed linear space.

(i) $\{x_k\}_{k\geqslant 1}$ *is convergent;*

(ii) $\{x_k\}_{k\geqslant 1}$ *is a Cauchy sequence, i.e., for any* $\varepsilon > 0$, *there exists a* $K \geqslant 1$ *(depending on* $\varepsilon > 0$) *such that*

$$\|x_k - x_\ell\|_2 < \varepsilon, \qquad \forall k, \ell \geqslant K.$$

(iii) *Every subsequence of* $\{x_k\}_{k\geqslant 1}$ *is convergent.*

(iv) *For each* $i = 1, 2, \cdots, n$, *the sequence* $\{x_k^i\}_{k\geqslant 1}$ *is convergent, where* $x_k = (x_k^1, \cdots, x_k^n)^\top$.

For a sequence $\{x_k\}_{k\geqslant 1}$, a *subsequence*, denoted by $\{x_{k_j}\}_{j\geqslant 1}$, is a sequence whose terms are selected from $\{x_k\}_{k\geqslant 1}$. For example, one can choose $k_j = 2j$, then $\{x_{k_j}\}_{j\geqslant 1} = \{x_{2j}\}_{j\geqslant 1} = \{x_2, x_4, \cdots\}$. In general, for any strictly increasing map $\sigma : \{1, 2, \cdots\} \to \{1, 2, \cdots\}$, $\{x_{\sigma(j)}\}_{j\geqslant 1}$ gives a subsequence of $\{x_k\}_{k\geqslant 1}$.

Note that by the above result, if in a sequence $\{x_k\}_{k\geqslant 1}$, one can find two subsequences which converge to two different limits, then the original sequence is not convergent. An easy example is $x_k = (-1)^k$, $k \geqslant 1$. Also, one sees that if a sequence $\{x_k\}_{k\geqslant 1}$ is convergent, the limit must be unique.

The following result is useful when one wants to prove a set to be closed.

Proposition 1.11. *Let* $G \subseteq \mathbb{R}^n$. *Then* G *is closed if and only if for any convergent sequence* $\{x_k\} \subseteq G$, *the limit* $\bar{x} \in G$.

Now, let us say a little bit more on the case $n = 1$. Recall that on $\mathbb{R}$, we have an *order* determined by "$\geqslant$" (or "$\leqslant$"). This order enables us to define the following notion which will play a fundamental role in the optimization theory.

Definition 1.12. *Let* $G \subset \mathbb{R}$.

(i) *A real number* $a \triangleq \inf G \in \mathbb{R}$ *is called the* infimum *(also called the* greatest lower bound*) of* G *if*

$$\begin{cases} x \geqslant a, & \forall x \in G, \\ \forall \varepsilon > 0, & \exists x \in G, \qquad \text{such that} \quad x < a + \varepsilon. \end{cases}$$

In the case that $\inf G \in G$, we call $\inf G$ the *minimum* of G, denote it by $\min G$.

(ii) A real number $b \triangleq \sup G \in \mathbb{R}$ is called the *supremum* (also called the *least upper bound*) of G if

$$\begin{cases} x \leqslant b, & \forall x \in G, \\ \forall \varepsilon > 0, & \exists x \in G, \quad \text{such that} \quad x > b - \varepsilon. \end{cases}$$

In the case that $\sup G \in G$, we call $\sup G$ the *maximum* of G, denote it by $\max G$.

We have the following basic result concerning the supremum and the infimum.

Proposition 1.13. *Let $G \subset \mathbb{R}$ be bounded. Then $\inf G$ and $\sup G$ uniquely exist. Moreover, there are sequences $\{x_k\}_{k \geqslant 1}, \{y_k\}_{k \geqslant 1} \subseteq G$ such that*

$$\lim_{k \to \infty} x_k = \inf G, \qquad \lim_{k \to \infty} y_k = \sup G.$$

Further, if G is closed (and hence it is compact), then

$$\sup G = \max G \in G, \qquad \inf G = \min G \in G.$$

In the above case, we call $\{x_k\}_{k \geqslant 1}$ and $\{y_k\}_{k \geqslant 1}$ a *minimizing sequence* and a *maximizing sequence*, respectively. The proof of the existence of the supremum and infimum is much more involved, for which we prefer not to get into details. However, the proof of the existence of maximizing and minimizing sequences easily follows the definition of the supremum and infimum, together with the following proposition.

Proposition 1.14. *Let $\{x_k\}_{k \geqslant 1} \subset \mathbb{R}$ be a sequence.*

(i) *If x_k is bounded from below and non-increasing, i.e. for some $M > 0$,*

$$x_k \geqslant -M, \qquad x_k \geqslant x_{k+1}, \quad \forall k \geqslant 1,$$

then x_k converges to $\inf_{k \geqslant 1} x_k$.

(ii) *If x_k is bounded from above and non-decreasing, i.e. for some $M > 0$,*

$$x_k \leqslant M, \qquad x_k \leqslant x_{k+1}, \quad \forall k \geqslant 1,$$

then x_k converges to $\sup_{k \geqslant 1} x_k$.

The following result characterizes compact sets.

Proposition 1.15. (Bolzano–Weierstrass) *Let $G \subset \mathbb{R}^n$. Then G is compact, i.e., G is bounded and closed, if and only if any sequence $\{x_k\}_{k \geqslant 1} \subseteq G$ admits a subsequence which is convergent to some $\bar{x} \in G$.*

Now, let $\{x_k\}_{k\geqslant 1}$ be a bounded sequence in $\mathbb{R}$. For any $\ell\geqslant 1$, the "tail sequence" $\{x_k\}_{k\geqslant\ell}$ of $\{x_k\}_{k\geqslant 1}$ is also bounded. Thus, by Proposition 1.13,

$$y_\ell \triangleq \inf_{k\geqslant\ell} x_k, \qquad z_\ell \triangleq \sup_{k\geqslant\ell} x_k$$

exist. Moreover, it is ready to see that $\{y_\ell\}_{\ell\geqslant 1}$ is bounded and non-decreasing, and $\{z_\ell\}_{\ell\geqslant 1}$ is bounded and non-increasing. Therefore, by Proposition 1.14, the following limits exist:

$$\underline{\lim_{k\to\infty}}\; x_k \triangleq \lim_{\ell\to\infty} y_\ell \equiv \lim_{\ell\to\infty} \inf_{k\geqslant\ell} x_k,$$

$$\overline{\lim_{k\to\infty}}\; x_k \triangleq \lim_{\ell\to\infty} z_\ell \equiv \lim_{\ell\to\infty} \sup_{k\geqslant\ell} x_k.$$

They are called the *limit inferior* (liminf, for short) and *limit superior* (limsup, for short) of sequence $\{x_k\}_{k\geqslant 1}$, respectively. We have the following simple and interesting result.

Proposition 1.16. *For any bounded sequence $\{x_k\}_{k\geqslant 1} \subset \mathbb{R}$, $\underline{\lim\limits_{k\to\infty}}\; x_k$ and $\overline{\lim\limits_{k\to\infty}}\; x_k$ exist. Moreover, if $\bar{x}$ is any limit point of $\{x_k\}_{k\geqslant 1}$ (i.e., there exists a subsequence $\{x_{k_j}\}_{j\geqslant 1}$ of $\{x_k\}_{k\geqslant 1}$ such that $x_{k_j} \to \bar{x}$), then*

$$\underline{\lim_{k\to\infty}}\; x_k \leqslant \bar{x} \leqslant \overline{\lim_{k\to\infty}}\; x_k.$$

Further,

$$\lim_{k\to\infty} x_k \text{ exists} \quad\Longleftrightarrow\quad \underline{\lim_{k\to\infty}}\; x_k = \overline{\lim_{k\to\infty}}\; x_k.$$

From the above, we see that for any bounded sequence $\{x_k\}_{k\geqslant 1}$, if K is the set of all limit points of the sequence, then

$$\underline{\lim_{k\to\infty}}\; x_k = \min K, \qquad \overline{\lim_{k\to\infty}}\; x_k = \max K.$$

Exercises

1. Let A and B be given by the following. Calculate $A \cup B$, $A \cap B$, $A \setminus B$ and $B \setminus A$:

(i) $A = [0,2]$ and $B = [0,1)$;

(ii) $A = \{1,2,3\}$, $B = \{a,b,c\}$;

(iii) $A = \{x \in \mathbb{R}^2 \mid x \geqslant 0\}$, $B = \{x \in \mathbb{R}^2 \mid x > 0\}$;

(iv) $A = \{a + b\sqrt{2} \mid a,b \in \mathbb{Q}\}$, $B = \{a + b\sqrt{3} \mid a,b \in \mathbb{Q}\}$, where $\mathbb{Q}$ is the set of all rational numbers.

2. Prove the following:

$$\left(\bigcap_{\lambda\in\Lambda}A_\lambda\right)^c=\bigcup_{\lambda\in\Lambda}A_\lambda^c.$$

3^*. Let

$$A=\{a+b\sqrt{2}\mid a,b\in\mathbb{Q}\},\quad B=\left\{\frac{a+b\sqrt{2}}{c+d\sqrt{2}}\mid a,b,c,d\in\mathbb{Q},c+d\sqrt{2}\neq0\right\}.$$

Prove or disprove $A=B$.

4. Let $\xi_1,\cdots,\xi_k\in\mathbb{R}^n$. Show that

$$\operatorname{span}\{\xi_1,\cdots,\xi_k\}\triangleq\left\{\sum_{i=1}^k\alpha_i\xi_i\mid\alpha_i\in\mathbb{R},\ 1\leqslant i\leqslant k\right\}$$

is a linear subspace of $\mathbb{R}^n$.

5. Let $\xi_1=(0,1,1)^\top$ and $\xi_2=(1,0,0)^\top$. Find $\operatorname{span}\{\xi_1,\xi_2\}$.

6. Determine linear independence of each given set of vectors:

(i) $\xi_1=(1,1,0)^\top$, $\xi_2=(1,0,1)^\top$, $\xi_3=(0,1,1)^\top$;

(ii) $\eta_1=(1,1,1)^\top$, $\eta_2=(2,0,2)^\top$, $\eta_3=(0,-1,0)^\top$.

7^*. Define

$$\|x\|_p=\begin{cases}\left(\displaystyle\sum_{i=1}^n|x_i|^p\right)^{\frac{1}{p}},&1\leqslant p<\infty,\\[2mm]\|x\|_\infty=\displaystyle\max_{1\leqslant i\leqslant n}|x_i|,&\end{cases}\qquad\forall x\in\mathbb{R}^n.$$

Show that $\|\cdot\|_p$, $1\leqslant p\leqslant\infty$, are norms on $\mathbb{R}^n$.

8^*. Let $\|\cdot\|_p$ be defined as above. Show that for any $p_1,p_2\in[1,\infty]$, there is a constant $K(p_1,p_2)$ such that

$$\|x\|_{p_1}\leqslant K(p_1,p_2)\|x\|_{p_2},\qquad\forall x\in\mathbb{R}^n.$$

What can you say about this result?

9^*. Prove Proposition 1.16.

10. For $G\subseteq\mathbb{R}^n$ defined below, find $\bar{G}$ and G°:

(i) $n=1$, $G=\{1,2,3,4\}$;

(ii) $n=2$, $G=\{(x,y)\mid x^2+y^2=1\}$;

(iii) $n=3$, $G=\{(x,y,z)\in\mathbb{R}^3\mid x^2+y^2+z^2<1\}$.

11. Define sequence $\{x_k\}_{k \geqslant 1}$ as below. Calculation $\varliminf\limits_{k \to \infty} x_k$ and $\varlimsup\limits_{k \to \infty} x_k$. Identify if the sequence is convergent, given your reason(s):

(i) $x_k = \frac{(-1)^k}{k}$, $k \geqslant 1$;

(ii) $x_k = 2 + (-1)^k$, $k \geqslant 1$;

(iii) $x_k = \sin \frac{k\pi}{2}$, $k \geqslant 1$.

12. Let G be given by the following. Find supremum, infimum, maximum and minimum.

(i) $G = \{x \in [0, 1] \mid x \text{ is irrational}\}$;

(ii) $G = \{\frac{1}{k} \mid k \in \mathbb{N}\}$;

(iii) $G = \{1 + \frac{(-1)^k}{k} \mid k \in \mathbb{N}\}$;

(iv) $G = \{x \in [0, 2\pi] \mid \sin x > \frac{1}{2}\}$.

2 Some Results of Linear Algebra

In this section, we briefly present some results from linear algebra. First of all, we recall that an $(m \times n)$ *matrix* A is defined to be the following arrays:

$$A = \begin{pmatrix} a_{11} & a_{12} & \cdots & a_{1n} \\ a_{21} & a_{22} & \cdots & a_{2n} \\ \vdots & \vdots & \cdots & \vdots \\ a_{m1} & a_{m2} & \cdots & a_{mn} \end{pmatrix}. \tag{2.1}$$

We also call the above A as a matrix of *order* $(m \times n)$. Sometimes, we use notation $A = (a_{ij})$ when the order is clear. The set of all $(m \times n)$ matrices is denoted by $\mathbb{R}^{m \times n}$. Two matrices $A = (a_{ij})$ and $B = (b_{ij})$ are equal if and only if they have the same order and $a_{ij} = b_{ij}$ for all i and j within their ranges. When $m = n$, A is called a *square matrix*. There are two special matrices: *identity matrix* I and *zero matrix* 0 which are defined by the following:

$$I = \begin{pmatrix} 1 & 0 & \cdots & 0 \\ 0 & 1 & \cdots & 0 \\ \vdots & \vdots & \ddots & \vdots \\ 0 & 0 & \cdots & 1 \end{pmatrix}, \qquad 0 = \begin{pmatrix} 0 & 0 & \cdots & 0 \\ 0 & 0 & \cdots & 0 \\ \vdots & \vdots & \ddots & \vdots \\ 0 & 0 & \cdots & 0 \end{pmatrix}.$$

Sometimes, we use I_n to represent the $(n \times n)$ identity matrix, indicating the size of the matrix. Note that an identity matrix must be square. But, a zero matrix does not have to be square.

On $\mathbb{R}^{m \times n}$, we define addition and scalar multiplication as follows: If $A = (a_{ij}), B = (b_{ij}) \in \mathbb{R}^{m \times n}$, and $\lambda \in \mathbb{R}$, then

$$A + B = (a_{ij} + b_{ij}), \qquad \lambda A = (\lambda a_{ij}). \tag{2.2}$$

One can show that $\mathbb{R}^{m \times n}$ is a linear space under the above defined addition and scalar multiplication (see Definition 1.2). Obviously, $\mathbb{R}^{m \times n}$ is quite different from $\mathbb{R}^n$. But, as far as linear spaces are concerned, they have the same nature. If for any $1 \leqslant i \leqslant m$, $1 \leqslant j \leqslant n$, we let $E(i,j) \in \mathbb{R}^{m \times n}$ be the matrix that the entry at (i,j) position is 1 and all other entries are zero, then $\{E(i,j) \mid 1 \leqslant i \leqslant m, 1 \leqslant j \leqslant n\}$ form a basis of $\mathbb{R}^{m \times n}$. Hence,

$$\dim \mathbb{R}^{m \times n} = mn.$$

Actually, $E(i,j)$ $(1 \leqslant i \leqslant m, 1 \leqslant j \leqslant n)$ are linearly independent, and

$$A \equiv (a_{ij}) = \sum_{i=1}^{m} \sum_{j=1}^{n} a_{ij} E(i,j), \qquad \forall A \equiv (a_{ij}) \in \mathbb{R}^{m \times n}.$$

In what follows, $E(i,j)$ might be defined in different $\mathbb{R}^{m \times n}$ and whose order $(m \times n)$ can be identified from the context.

Next, for $A = (a_{ij}) \in \mathbb{R}^{m \times n}$ given by (2.1), we define its *transpose* $A^\top \in \mathbb{R}^{n \times m}$ by the following:

$$A^\top = \begin{pmatrix} a_{11} & a_{21} & \cdots & a_{m1} \\ a_{12} & a_{22} & \cdots & a_{m2} \\ \vdots & \vdots & \cdots & \vdots \\ a_{1n} & a_{1n} & \cdots & a_{mn} \end{pmatrix}. \tag{2.3}$$

Thus, A and $A^\top$ have different orders unless $m = n$. When $A \in \mathbb{R}^{n \times n}$, and $A^\top = A$, it is called a *symmetric* matrix. The set of all symmetric matrices of order $(n \times n)$ is denoted by $\mathbb{S}^n$. Also, for any square matrix $A = (a_{ij}) \in \mathbb{R}^{n \times n}$, we define

$$\operatorname{tr}(A) = \sum_{i=1}^{n} a_{ii}, \tag{2.4}$$

which is called the *trace* of A.

Any $A = (a_{ij}) \in \mathbb{R}^{m \times n}$ has n columns which are called the *column vectors* of A (all of them are in $\mathbb{R}^m$). Likewise, A has m rows which are called the *row vectors* of A (whose transpose are the column vectors of $A^\top$ and belong to $\mathbb{R}^n$). On the other hand, any vector in $\mathbb{R}^n$ can be regarded as an $(n \times 1)$ matrix, and any m vectors in $\mathbb{R}^n$ can form an $(n \times m)$ matrix. Such an observation is very useful.

Next, for $A \in \mathbb{R}^{m \times n}$ and $B \in \mathbb{R}^{n \times \ell}$ given by

$$A = \begin{pmatrix} a_{11} & a_{12} & \cdots & a_{1n} \\ a_{21} & a_{22} & \cdots & a_{2n} \\ \vdots & \vdots & \cdots & \vdots \\ a_{m1} & a_{m2} & \cdots & a_{mn} \end{pmatrix}, \qquad B = \begin{pmatrix} b_{11} & b_{12} & \cdots & b_{1\ell} \\ b_{21} & b_{22} & \cdots & b_{2\ell} \\ \vdots & \vdots & \cdots & \vdots \\ b_{n1} & b_{n2} & \cdots & b_{n\ell} \end{pmatrix},$$

we define the product $AB = (c_{ij}) \in \mathbb{R}^{m \times \ell}$ of A and B with the entries c_{ij} given by the following:

$$c_{ij} = \sum_{k=1}^{n} a_{ik} b_{kj}, \qquad 1 \leqslant i \leqslant m, \ 1 \leqslant j \leqslant \ell.$$

In particular, regarding any $x, y \in \mathbb{R}^n$ as $x, y \in \mathbb{R}^{n \times 1}$, we see that

$$x^\top y = \sum_{k=1}^{n} x_k y_k = x \cdot y.$$

Hereafter, in $\mathbb{R}^n$, we will identify $x \cdot y$ with $x^\top y$.

Note that in order the product AB makes sense, the number of columns of A should be equal to the number of rows of B. Thus, when AB is defined, BA is not necessarily define. Further, if both AB and BA are all defined, they may have different orders, and even if AB and BA have the same order, we may still not necessarily have $AB = BA$. However, it is interesting that as long as AB is defined, $B^\top A^\top$ will also be defined and

$$(AB)^\top = B^\top A^\top.$$

In the case $A \in \mathbb{R}^{n \times n}$, sometimes there exists a $B \in \mathbb{R}^{n \times n}$ such that

$$AB = BA = I.$$

When this happens, we say that A is *invertible* (or *non-singular*), and in this case, one can show that the matrix B satisfying the above must be unique. Hence, we denote such a B by A^{-1} which is called the *inverse* of A. We have the following result concerning the inverse of matrices.

Proposition 2.1. (i) *Let* $A \in \mathbb{R}^{n \times n}$ *be invertible. Then so is* $A^\top$ *and*

$$(A^\top)^{-1} = (A^{-1})^\top.$$

(ii) *Let* $A, B \in \mathbb{R}^{n \times n}$ *be invertible. Then so are* AB *and* BA, *and*

$$(AB)^{-1} = B^{-1} A^{-1}.$$

We note that for any $A, B \in \mathbb{R}^{m \times n}$, the product $AB^\top \in \mathbb{R}^{m \times m}$ and $A^\top B \in \mathbb{R}^{n \times n}$ are always defined. Further the following proposition holds.

Proposition 2.2. (i) *Let $A \in \mathbb{R}^{m \times n}$ and $B \in \mathbb{R}^{n \times m}$. Then*

$$\operatorname{tr}[AB] = \operatorname{tr}[BA]. \tag{2.5}$$

(ii) *The map $(A, B) \mapsto \operatorname{tr}[A^\top B]$ is an inner product on $\mathbb{R}^{m \times n}$. Consequently,*

$$|A| = \left(\operatorname{tr}[A^\top A] \right)^{\frac{1}{2}}, \qquad \forall A \in \mathbb{R}^{m \times n}, \tag{2.6}$$

defines a norm in $\mathbb{R}^{m \times n}$.

From the above, we see that under $(A, B) \mapsto \operatorname{tr}[A^\top B]$, $\mathbb{R}^{m \times n}$ becomes an inner product space, and therefore, it is also a normed linear space. Further, one can directly check that $\{E(i, j) \mid 1 \leqslant i \leqslant m, 1 \leqslant j \leqslant n\}$ forms an orthonormal basis of $\mathbb{R}^{m \times n}$, under the above-defined inner product.

Next, for any $A \in \mathbb{R}^{m \times n}$, we introduce three *elementary row operations*:

(i) Exchange the i-th and the j-th rows;

(ii) Multiply the i-th row by $\lambda \in \mathbb{R}$, $\lambda \neq 0$; and

(iii) Add the μ multiple of the j-th row to the i-th row, $\mu \in \mathbb{R}$.

We now translate these three operations into products of matrices. To this end, we recall the definition of $E(i, j) \in \mathbb{R}^{m \times m}$, $1 \leqslant i, j \leqslant m$. Define three types of *elementary matrices* as follows:

$$\begin{cases} T_1(i, j) = \sum_{k \neq i, j} E(k, k) + E(i, j) + E(j, i), & i \neq j, \\ T_2(i \, ; \lambda) = \sum_{k \neq i} E(k, k) + \lambda E(i, i), & \lambda \neq 0, \\ T_3(i, j \, ; \mu) = I_m + \mu E(i, j), & i \neq j. \end{cases}$$

We see that $T_1(i, j)$ is the matrix obtained from I_m by exchange the i-th and the j-th columns; $T_2(i \, ; \lambda)$ is the matrix obtained from I_m by replacing the i-th diagonal entry 1 by λ; and $T_3(i, j \, ; \mu)$ is the matrix obtained from I_m by replacing the 0 entry at (i, j) position by $\mu \in \mathbb{R}$. Direct computations show that the products $T_1(i, j)A$, $T_2(i \, ; \lambda)A$, and $T_3(i, j \, ; \mu)A$ are exactly the three elementary row operations applied to A, respectively.

Likewise, we may introduce *elementary column operations*, which can be achieved as follows: First make a transpose to the matrix, followed by an elementary row operation, then make a transpose back. We leave the details to the interested readers.

Now, for a given $A = (a_{ij}) \in \mathbb{R}^{m \times n}$, let $1 \leqslant i_1 < \cdots < i_k \leqslant m$ and $1 \leqslant j_1 < \cdots < j_\ell \leqslant n$. Then $B = (b_{\alpha\beta}) \in \mathbb{R}^{k \times \ell}$ is called a *sub-matrix* of A where

$$b_{\alpha\beta} = a_{i_\alpha j_\beta}, \qquad 1 \leqslant \alpha \leqslant k, \ 1 \leqslant \beta \leqslant \ell.$$

When $k = \ell$, we call the above B a *square sub-matrix* of A. We now introduce an important notion for square matrices, which will be useful below. For any $A = (a_{ij}) \in \mathbb{R}^{n \times n}$ we define its *determinant* $\det(A)$ by the following (inductively):

$$\det(A) = \begin{cases} a_{11}, & n = 1, \\ \sum_{j=1}^{n} (-1)^{i+j} a_{ij} \det(A_{ij}), & n > 1, \end{cases} \tag{2.7}$$

where $A_{ij} \in \mathbb{R}^{(n-1) \times (n-1)}$ is the sub-matrix of A obtained by deleting the i-th row and the j-th column of A. We also define

$$\mathrm{adj}\,(A) = \left((-1)^{i+j} \det(A_{j\,i}) \right).$$

This is called the *adjugate* of A. One should note that the entry of $\mathrm{adj}\,(A)$ at (i, j) is $(-1)^{i+j} \det(A_{j\,i})$ not $(-1)^{i+j} \det(A_{ij})$. The following proposition collects basic properties of determinants.

Proposition 2.3. *Let $A, B \in \mathbb{R}^{n \times n}$. Then*

(i) $\det(A^\top) = \det(A)$;

(ii) *If two rows or two columns of A are the same, then $\det(A) = 0$;*

(iii) *If B is obtained by exchanging two rows (or two columns) of A, then $\det(B) = -\det(A)$;*

(iv) *If B is obtained by adding a multiple of a row (or a column) to another row (column) of A, then $\det(B) = \det(A)$.*

(v) $\det(AB) = \det(A)\det(B)$.

(vi) $A \cdot \mathrm{adj}(A) = \det(A)I$.

The following result presents a very interesting usage of determinant.

Proposition 2.4. *Let $\xi_1, \cdots, \xi_k \in \mathbb{R}^n$ and let*

$$A = (\xi_1, \cdots, \xi_k) \in \mathbb{R}^{n \times k}.$$

Then $\xi_1, \cdots, \xi_k$ are linearly independent if and only if there exists a $(k \times k)$ square sub-matrix B of A such that $\det(B) \neq 0$. In particular, if $k = n$, then the column vectors of A are linearly independent if and only if $\det(A) \neq 0$. In this case, A must be invertible and

$$A^{-1} = \frac{1}{\det(A)} \, \mathrm{adj}\,(A). \tag{2.8}$$

By some straightforward calculations, we have

$$\det \big(T_1(i,j)\big) = -1, \quad \det \big(T_2(i\,;\lambda)\big) = \lambda, \quad \det \big(T_3(i,j\,;\mu)\big) = 1. \tag{2.9}$$

Therefore, by Proposition 2.4, $T_1(i,j)$, $T_2(i\,;\lambda)$ and $T_3(i,j\,;\mu)$ are all invertible, for any $\lambda \neq 0$ and $\mu \in \mathbb{R}$. Actually, one has

$$\begin{aligned}
T_1(i,j)^{-1} &= T_1(i,j), \\
T_2(i\,;\lambda)^{-1} &= T_2(i\,;\lambda^{-1}), \\
T_3(i,j\,;\mu)^{-1} &= T_3(i,j\,;-\mu),
\end{aligned}$$

which are still elementary matrices.

Now, we return to the product of matrices. As an important special case, for any $A = (a_{ij}) \in \mathbb{R}^{m \times n}$ and $x = (x_1, \cdots, x_n)^\top \in \mathbb{R}^n$, Ax makes sense:

$$Ax = \begin{pmatrix} a_{11} & a_{12} & \cdots & a_{1n} \\ a_{21} & a_{22} & \cdots & a_{2n} \\ \vdots & \vdots & \cdots & \vdots \\ a_{m1} & a_{m2} & \cdots & a_{mn} \end{pmatrix} \begin{pmatrix} x_1 \\ x_2 \\ \vdots \\ x_n \end{pmatrix} = \begin{pmatrix} b_1 \\ b_2 \\ \vdots \\ b_m \end{pmatrix} = b \in \mathbb{R}^m,$$

with

$$b_j = \sum_{k=1}^{n} a_{ik} x_k, \quad 1 \leqslant j \leqslant m.$$

Thus, matrix $A \in \mathbb{R}^{m \times n}$ induces a map from $\mathbb{R}^n$ to $\mathbb{R}^m$. Moreover, this map has the following property:

$$A(\lambda x + \mu y) = \lambda Ax + \mu Ay, \qquad \forall x, y \in \mathbb{R}^n, \; \lambda, \mu \in \mathbb{R}. \tag{2.10}$$

The above property is referred to as the *linearity*. Hence, we say that the map induced by A (hereafter, it will be identified with A) is a *linear map* (or a *linear transformation*). Further, if we write

$$A = (\xi_1, \cdots, \xi_n), \qquad \xi_i \in \mathbb{R}^m, \; 1 \leqslant i \leqslant n,$$

and $x = (x_1, \cdots, x_n)^\top \in \mathbb{R}^n$, then

$$Ax = \sum_{i=1}^{n} x_i \xi_i,$$

which is a linear combination of $\xi_1, \cdots, \xi_n$. We now introduce some more relevant notions.

- The *kernel* $\mathcal{N}(A)$ of a matrix $A \in \mathbb{R}^{m \times n}$ is defined by

$$\mathcal{N}(A) = \Big\{ x \in \mathbb{R}^n \mid Ax = 0 \Big\}. \tag{2.11}$$

- The *range* $\mathcal{R}(A)$ of a matrix $A \in \mathbb{R}^{m \times n}$ is defined by

$$\mathcal{R}(A) = \Big\{ Ax \mid x \in \mathbb{R}^n \Big\}. \tag{2.12}$$

We have the following result.

Proposition 2.5. (i) *Let $A \in \mathbb{R}^{m \times n}$. Then $\mathcal{N}(A)$ and $\mathcal{R}(A)$ are linear subspaces of $\mathbb{R}^n$ and $\mathbb{R}^m$, respectively. Moreover,*

$$\mathcal{R}(A)^\perp = \mathcal{N}(A^\top), \qquad \mathcal{N}(A)^\perp = \mathcal{R}(A^\top). \tag{2.13}$$

(ii) *Let $A \in \mathbb{R}^{n \times n}$. Then*

$$A^{-1} \text{ exists} \iff \mathcal{N}(A) = \{0\} \iff \mathcal{R}(A) = \mathbb{R}^n \iff \det(A) \neq 0$$
$$\iff A = E_1 E_2, \cdots E_k, \text{ with } E_\ell \ (1 \leqslant \ell \leqslant k) \text{ being elementary matrices.}$$

Proof. Let us prove (2.13). Note that

$$y \in \mathcal{R}(A)^\perp \iff 0 = y^\top (Ax) = (A^\top y)^\top x, \quad \forall x \in \mathbb{R}^n$$
$$\iff A^\top y = 0 \iff y \in \mathcal{N}(A^\top).$$

This proves the first relation in (2.13).

Applying the first relation in (2.13) to $A^\top$, we have

$$\mathcal{R}(A^\top)^\perp = \mathcal{N}(A).$$

Then taking orthogonal complimentary on both sides, we obtain the second relation in (2.13).

The proofs of the other conclusions are left to the readers. $\qquad \square$

Next, for any $A = (a_1, \cdots, a_n) \in \mathbb{R}^{m \times n}$ with $a_i \in \mathbb{R}^m$, we define the *rank* of A by the following:

$$\text{rank}\,(A) = \dim \mathcal{R}(A) = \dim \Big(\text{span}\,\{a_1, \cdots, a_n\} \Big). \tag{2.14}$$

Thus, rank (A) is the maximum number of linearly independent vectors in the set $\{a_1, \cdots, a_n\}$. According to Proposition 2.4, this is also the maximum order of the square sub-matrix B of A so that $\det(B) \neq 0$. Such an observation leads to a proof of the following proposition, part (i).

Proposition 2.6. (i) *For any* $A \in \mathbb{R}^{m \times n}$,

$$\text{rank}\,(A) = \text{rank}\,(A^\top). \tag{2.15}$$

(ii) *Let* $B \in \mathbb{R}^{m \times m}$ *and* $C \in \mathbb{R}^{n \times n}$ *be invertible. Then for any* $A \in \mathbb{R}^{m \times n}$,

$$\text{rank}\,(BA) = \text{rank}\,(AC) = \text{rank}\,(A). \tag{2.16}$$

In particular, for any $\lambda \neq 0$, $\mu \in \mathbb{R}$,

$$\begin{aligned}
\text{rank}\,\big(T_1(i,j)A\big) &= \text{rank}\,\big(T_2(i\,;\lambda)A\big) \\
&= \text{rank}\,\big(T_3(i,j\,;\mu)A\big) = \text{rank}\,(A).
\end{aligned} \tag{2.17}$$

One can usually use the above to find the rank of a given matrix A by performing elementary row operations. Also, this can be used to find the inverse of A as follows: Assume A is invertible. Then by (ii) of Proposition 2.5,

$$A = E_1 E_2 \cdots E_k,$$

with $E_1, \cdots, E_k$ being elementary matrices. This leads to

$$A^{-1} = E_k^{-1} E_{k-1}^{-1} \cdots E_2^{-1} E_1^{-1}.$$

Hence, by performing elementary row operations $E_1^{-1}, E_2^{-1}, \cdots, E_{k-1}^{-1}, E_k^{-1}$ consecutively to the *augmented matrix* (A, I), we have

$$E_1^{-1} E_2^{-1} \cdots E_k^{-1}(A, I) = (A^{-1}A, A^{-1}) = (I, A^{-1}).$$

Practically, the above method is very effective.

The following result is about *linear equation* of form $Ax = b$.

Proposition 2.7. *Let* $A \in \mathbb{R}^{m \times n}$.

(i) *Let* $b \in \mathbb{R}^m$. *Linear equation* $Ax = b$ *has at least one solution if and only if* $b \in \mathcal{R}(A)$, *which is also equivalent to the following:*

$$\text{rank}\,(A) = \text{rank}\,(A, b). \tag{2.18}$$

(ii) *Linear equation* $Ax = b$ *has at least one solution for every* $b \in \mathbb{R}^m$ *if and only if* $\mathcal{R}(A) = \mathbb{R}^m$.

(iii) *Linear equation $Ax = b$ has at most one solution for some $b \in \mathbb{R}^m$ if and only if $\mathcal{N}(A) = \{0\}$, which is also equivalent to the fact that the equation $Ax = 0$ only admits the zero solution. If this happens, then it is necessary that $m \geqslant n$.*

(iv) *Linear equation $Ax = b$ admits a unique solution for any $b \in \mathbb{R}^m$ if and only if $m = n$ and $\det(A) \neq 0$; in this case, A^{-1} exists and for any $b \in \mathbb{R}^n$, the unique solution of equation $Ax = b$ admits the following representation: $x = A^{-1}b$.*

From the above, we obtain a very important corollary.

Corollary 2.8. *Let $A \in \mathbb{R}^{m \times n}$. Linear equation $Ax = 0$ admits a non-zero solution if and only if either $m < n$, or $m = n$ with $\det(A) = 0$.*

We also note that by performing the elementary row operations, one can solve linear equation $Ax = b$. More precisely, suppose A is invertible. Then with elementary row operations, say, $E_1, E_2, \cdots, E_k$, one will end up with

$$E_k E_{k-1} \cdots E_1 A = I.$$

Therefore, applying such a sequence of operations, we will have

$$x = E_k E_{k-1} \cdots E_1 A x = E_k E_{k-1} \cdots E_1 b \equiv A^{-1}b,$$

which gives the solution. This procedure is called the *Gaussian elimination method*. Even when A is not invertible, this method will still work. In general, after a sequence of elementary row operations, we end up with an equation $A_0 x = b_0$ with A_0 being a simplified form, called a *reduced echelon form* of A. Since elementary row operations are reversible, the resulting equation $A_0 x = b_0$ is equivalent to the original one, in the sense that they have the same set of solutions.

Next, we introduce the following definition.

Definition 2.9. *Let $A \in \mathbb{R}^{n \times n}$.*

(i) *The polynomial $\det(\lambda I - A)$ is called the *characteristic polynomial* of A.*

(ii) *Any root of the *characteristic equation* $\det(\lambda I - A) = 0$, which could be complex, is called an *eigenvalue* of A. The set of all eigenvalues of A is denoted by $\sigma(A)$ and is called the *spectrum* of A.*

(iii) *If $\lambda \in \mathbb{C}$ is an eigenvalue of A and $\xi, \eta \in \mathbb{R}^n$ such that $\xi + i\eta \neq 0$ and $A(\xi + i\eta) = \lambda(\xi + i\eta)$, then $\xi + i\eta$ is called an *eigenvector* of A.*

In the above, $i = \sqrt{-1}$. Now, for any $A \in \mathbb{S}^n$, we define a map $x \mapsto x^\top A x$ which is called a *quadratic form*. This is a map from $\mathbb{R}^n$ to $\mathbb{R}$. We introduce the following notions.

Definition 2.10. Let $A \in \mathbb{S}^n$.

(i) A is said to be *positive semi-definite*, denoted by $A \geqslant 0$, if

$$x^\top A x \geqslant 0, \qquad \forall x \in \mathbb{R}^n.$$

(ii) A is said to be *positive definite*, denoted by $A > 0$, if

$$x^\top A x > 0, \qquad \forall x \in \mathbb{R}^n \setminus \{0\}.$$

(iii) A is said to be *negative semi-definite*, denoted by $A \leqslant 0$, if $-A$ is positive semi-definite; and A is said to be *negative definite*, denoted by $A < 0$, if $-A$ is positive definite.

We have the following criterion for positive definiteness of symmetric matrices.

Theorem 2.11. (Sylvester) *Let $A = (a_{ij}) \in \mathbb{S}^n$. Define*

$$d_1 = a_{11}, \quad d_2 = \det \begin{pmatrix} a_{11} & a_{12} \\ a_{21} & a_{22} \end{pmatrix}, \quad \cdots, \quad d_n = \det(A).$$

Then A is positive definite if and only if

$$d_1, d_2, \cdots, d_n > 0. \tag{2.19}$$

Note that we do not have a similar criterion for positive semi-definiteness of matrices. Namely, if instead of (2.19), one has

$$d_1, d_2, \cdots, d_n \geqslant 0,$$

the matrix A is not necessarily positive semi-definite. For example, for

$$A = \begin{pmatrix} 0 & 0 \\ 0 & -1 \end{pmatrix},$$

we have $d_1 = d_2 = 0$. But the above matrix A is not positive semi-definite. However, we have the following result.

Proposition 2.12. *Let $A \in \mathbb{S}^n$. Then $\sigma(A) \subseteq \mathbb{R}$. Moreover, A is positive semi-definite if and only if $\sigma(A) \subseteq [0, \infty)$, and A is positive definite if and only if $\sigma(A) \subseteq (0, \infty)$.*

Sometimes, the following extension of Definition 2.10 is useful.

Definition 2.13. Let $A \in \mathbb{S}^n$ and $X \subseteq \mathbb{R}^n$ be a subspace.

(i) A is said to be *positive semi-definite on X* if

$$x^\top A x \geqslant 0, \qquad \forall x \in X. \tag{2.20}$$

(ii) A is said to be *positive definite on X* if

$$x^\top A x > 0, \qquad \forall x \in X \setminus \{0\}. \tag{2.21}$$

(iii) A is said to be *negative semi-definite on X* if $-A$ is positive semi-definite on X; and A is said to be *negative definite on X* if $-A$ is positive definite on X.

Let us look at two typical cases of the above.

(i) $X = \mathcal{R}(B)$, for some $B \in \mathbb{R}^{n \times m}$. In this case, $x \in X$ if and only if $x = By$ for some $y \in \mathbb{R}^m$. Thus,

$$x^\top A x = y^\top (B^\top A B) y.$$

Consequently, A is positive semi-definite (resp. positive definite, negative semi-definite, negative definite) on X if and only if so is $B^\top A B$ on $\mathbb{R}^m$.

(ii) $X = \mathcal{N}(B)$, for some $B \in \mathbb{R}^{m \times n}$. In this case, we let $\mathcal{N}(B) = \text{span}\{\xi_1, \cdots, \xi_k\} \subseteq \mathbb{R}^n$ and let

$$C = (\xi_1, \cdots, \xi_k) \in \mathbb{R}^{n \times k}.$$

Then $X = \mathcal{R}(C)$. Thus, by (i), we see that A is positive semi-definite (resp. positive definite, negative semi-definite, negative definite) on X if and only if so is $C^\top A C$ on $\mathbb{R}^k$. Note that $\mathcal{N}(B)$ can be found by solving linear equation $Bx = 0$.

Exercises

1. Show that $\mathbb{R}^{m \times n}$ is a linear space under the addition and scalar multiplication defined by (2.2).

2. Show that $(A, B) \mapsto \text{tr}[A^\top B]$ is an inner product of $\mathbb{R}^{m \times n}$, and $\{E(i,j) \mid 1 \leqslant i \leqslant m, 1 \leqslant j \leqslant n\}$ is an orthonormal basis of $\mathbb{R}^{m \times n}$.

3. Calculate the following determinants:

(i) $\begin{vmatrix} 1 & 2 \\ 2 & 1 \end{vmatrix}$; (ii) $\begin{vmatrix} \cos\theta & \sin\theta \\ -\sin\theta & \cos\theta \end{vmatrix}$;

(iii) $\begin{vmatrix} 3 & 0 & 1 \\ 2 & 1 & 2 \\ 0 & 3 & 2 \end{vmatrix}$;   (iv) $\begin{vmatrix} \sin\varphi\cos\theta & \sin\varphi\sin\theta & \cos\varphi \\ \rho\cos\varphi\cos\theta & \rho\cos\varphi\sin\theta & -\rho\sin\varphi \\ -\rho\sin\varphi\sin\theta & \rho\sin\varphi\cos\theta & 0 \end{vmatrix}$.

4*. Let $\xi_1, \cdots, \xi_k \in \mathbb{R}^n$ and let

$$A = (\xi_1, \cdots, \xi_k) \in \mathbb{R}^{n \times k}.$$

Show that $\xi_1, \cdots, \xi_k$ are linearly independent if and only if there exists a $(k \times k)$ square sub-matrix B of A such that $\det(B) \neq 0$.

5. Let $A \in \mathbb{R}^{m \times n}$. Show that $\mathcal{N}(A)$ and $\mathcal{R}(A)$ are linear subspaces of $\mathbb{R}^n$ and $\mathbb{R}^m$, respectively.

6. Let A be given by the following. Find $\mathcal{N}(A)$, $\mathcal{N}(A^\top)$, $\mathcal{R}(A)$, and $\mathcal{R}(A^\top)$:

(i) $A = \begin{pmatrix} 1 & 0 & 0 \\ 2 & 1 & 0 \end{pmatrix}$;   (ii) $A = \begin{pmatrix} 2 & 1 \\ 3 & 0 \\ 1 & 0 \\ 1 & 1 \end{pmatrix}$.

Compare $\mathcal{N}(A)$ with $\mathcal{R}(A^\top)^\perp$, and $\mathcal{N}(A^\top)$ with $\mathcal{R}(A)^\perp$.

7. Find examples showing that AB and BA have the same order, but $AB \neq BA$.

8*. Let $A \in \mathbb{R}^{n \times n}$ such that $AB = BA$ for all $B \in \mathbb{R}^{n \times n}$. Show that $A = \lambda I$ for some $\lambda \in \mathbb{R}$.

9. Calculate eigenvalues and corresponding eigenvectors of A:

(i) $A = \begin{pmatrix} 1 & 2 \\ -1 & 4 \end{pmatrix}$; (ii) $A = \begin{pmatrix} 1 & 2 \\ 2 & 1 \end{pmatrix}$.

10. Let $A \in \mathbb{S}^n$. Show that every eigenvalue of A must be a real number.

11. Let $A \in \mathbb{S}^n$ be positive definite. Define

$$\langle x, y \rangle_A = x^\top A y, \qquad x, y \in \mathbb{R}^n.$$

Show the above is an inner product in $\mathbb{R}^n$.

12. Let $A = \begin{pmatrix} 2 & 1 \\ 1 & -1 \end{pmatrix}$.

(i) Determine if A is positive definite.

(ii) Let $X = \text{span}\left\{ \begin{pmatrix} 1 \\ 1 \end{pmatrix} \right\}$. Is A positive definite on X? Give your reasons.

13*. For any $p \in [1, \infty]$, let

$$\|A\|_p = \sup_{\|x\|_p = 1} \|Ax\|_p, \qquad A \in \mathbb{R}^{m \times n}.$$

Show that $\| \cdot \|_p$ is a norm on $\mathbb{R}^{m \times n}$. (Recall $\|x\|_p$ defined by Problem 7 of Exercises in Section 1.)

3 Functions: Limits and Continuity

In what follows, for notational simplicity, $\| \cdot \|$ will stand for the Euclidean norm of $\mathbb{R}^n$ (unless otherwise stated). Any map $f : \mathbb{R}^n \to \mathbb{R}$ is called a *(scalar-valued) function*. If $n = 1$, such a function is called a *single-variable* function and if $n \geqslant 2$, such a function is called a *multi-variable* function. We make the following convention: For general case $n \geqslant 1$, the vectors in $\mathbb{R}^n$ are denoted by $x = (x_1, \cdots, x_n)^\top$, $y = (y_1, \cdots, y_n)^\top$, etc. However, traditionally, for $n = 1, 2, 3$, the vectors in $\mathbb{R}^n$ are denoted by x, $(x, y)^\top$, and $(x, y, z)^\top$, respectively. The following definition collects several relevant notions.

Definition 3.1. Let $G \subseteq \mathbb{R}^n$, and $f : G \to \mathbb{R}$.

(i) Let $\bar{x} \in \bar{G}$. If for any $\varepsilon > 0$, there exists a $\delta = \delta(\bar{x}, \varepsilon) > 0$, such that

$$|f(x) - L| < \varepsilon, \qquad \forall \|x - \bar{x}\| < \delta, \ x \in G \setminus \{\bar{x}\}, \qquad (3.1)$$

then we say that $f(x)$ admits a *limit* L as x approaches to $\bar{x}$, denoted by

$$\lim_{x \to \bar{x}} f(x) = L.$$

In the case that $\bar{x} \in G$ and $f(\bar{x}) = L$, we say that $f(\cdot)$ is *continuous* at $\bar{x}$. If $f(\cdot)$ is continuous at every point $\bar{x} \in G$, we say that $f(\cdot)$ is continuous on G. Further, if $f(\cdot)$ is continuous on G with $\delta(\bar{x}, \varepsilon)$ independent of $\bar{x}$, then $f(\cdot)$ is *uniformly continuous* on G.

(ii) Suppose for $\bar{x} \in G$, there exist $L, \delta > 0$ such that

$$|f(x) - f(y)| \leqslant L\|x - y\|, \qquad \forall x, y \in G \cap B_\delta(\bar{x}).$$

Then $f(\cdot)$ is said to be *Lipschitz continuous* near $\bar{x}$. If $f(\cdot)$ is Lipschitz continuous near any $x \in G$, we say that $f(\cdot)$ is *locally* Lipschitz continuous on G. Further, if $f(\cdot)$ is locally Lipschitz continuous on G with constant(s) L and δ uniformly in $x \in G$, then $f(\cdot)$ is said to be *globally* Lipschitz continuous on G.

Definition 3.2. Let $G \subseteq \mathbb{R}^n$ and $f : G \to \mathbb{R}$.

(i) A point $x_0 \in G$ is called a *global minimum* (respectively, *global maximum*) of $f(\cdot)$ over G if

$$f(x_0) \leqslant f(x), \quad \forall x \in G. \qquad \left(\text{resp. } f(x_0) \geqslant f(x), \quad \forall x \in G. \right)$$

When the above inequalities are strict for any $x \in G \setminus \{x_0\}$, we call x_0 a *strict global minimum* (respectively, *strict global maximum*) of $f(\cdot)$ over G.

(ii) A point $x_0 \in G$ is called a (*strict*) *local minimum* (respectively, (*strict*) *local maximum*) of $f(\cdot)$ if there exists a $\delta > 0$ such that x_0 is a (strict) global minimum (respectively, (strict) global maximum) of $f(\cdot)$ over $G \cap B_\delta(x_0)$.

Global minimum and global maximum are also called *absolute minimum* and *absolute maximum*, respectively. Further, both are simply called global or absolute *extremes*. Likewise, local minimum and local maximum are also called *relative minimum* and *relative maximum*, respectively. Both local minimum and local maximum are called *local extremes*. It is clear that if $x_0 \in G$ is a global minimum (respectively, global maximum) of $f(\cdot)$ over G, then

$$f(x_0) = \min f(G), \qquad \left(\text{resp. } f(x_0) = \max f(G), \right)$$

and if $x_0 \in G$ is a local minimum (respectively, local maximum), then for some $\delta > 0$,

$$f(x_0) = \min f\big(G \cap B_\delta(x_0)\big). \qquad \left(\text{resp. } f(x_0) = \max f\big(G \cap B_\delta(x_0)\big). \right)$$

The following result is concerned with the existence of extremes.

Theorem 3.3. *Let $G \subseteq \mathbb{R}^n$ and $f : G \to \mathbb{R}$ be continuous.*

(i) *If G is compact, so is $f(G) \subset \mathbb{R}$. Therefore,*

$$\inf f(G) = \min f(G) \in f(G), \quad \sup f(G) = \max f(G) \in f(G). \qquad (3.2)$$

(ii) *If G is connected, so is $f(G) \subseteq \mathbb{R}$.*

Proof. (i) We show that $f(G)$ is bounded and closed. Suppose $f(G)$ were not bounded. Then there would be a sequence $x_k \in G$ such that $|f(x_k)| \to \infty$. On the other hand, since G is compact, $\{x_k\}$ admits a convergent subsequence. We may assume that x_k itself is convergent to some $\bar{x} \in G$. Then by the continuity of $f(\cdot)$, we obtain

$$|f(\bar{x})| = \lim_{k \to \infty} |f(x_k)| = \infty,$$

which is a contradiction. Hence, $f(G)$ is bounded. To show the closedness of $f(G)$, let $y_k \in f(G)$ such that $y_k \to \bar{y}$ (as $k \to \infty$). We can find $x_k \in G$ such that $f(x_k) = y_k$. By the compactness of G, we may assume that $x_k \to \bar{x}$. Then by the continuity of $f(\cdot)$, one has $\bar{y} = f(\bar{x}) \in f(G)$. Hence, $f(G)$ is compact. Then (3.2) follows from Proposition 1.13.

(ii) Since $f(G) \subseteq \mathbb{R}$, to prove its connectedness, it suffices to prove that for any $x_1, x_2 \in G$ with $f(x_1) < f(x_2)$,

$$[f(x_1), f(x_2)] \subseteq f(G). \tag{3.3}$$

If this were not the case, then there would be some number, say, a, such that

$$f(x_1) < a < f(x_2), \quad a \notin f(G).$$

Define

$$G_1 = \{x \in G \mid f(x) < a\}, \quad G_2 = \{x \in G \mid f(x) > a\}.$$

Clearly, $G_1 \bigcup G_2 = G$. On the other hand, if there exists a point $\bar{x} \in \overline{G}_1 \bigcap G_2$, then $f(\bar{x}) > a$, and one must have a sequence $x_k \in G_1$ converging to $\bar{x}$. It follows from $x_k \in G_1$ that $f(x_k) < a$. Hence, by the continuity of $f(\cdot)$, we have

$$a < f(\bar{x}) = \lim_{k \to \infty} f(x_k) \leqslant a,$$

which is a contradiction. Thus, $\overline{G}_1 \bigcap G_2 = \varnothing$. Likewise, $G_1 \bigcap \overline{G}_2 = \varnothing$. Hence, G is disconnected, contradicting to the connectedness of G. □

Part (i) of the above theorem is called *Weierstrass Theorem*. Note that (3.2) means that when G is compact and $f : G \to \mathbb{R}$ is continuous, then $f(\cdot)$ attains its minimum and maximum over G. Part (ii) of the above is called the (generalized) *Intermediate Value Theorem*, which says that if $a, b \in f(G)$, then for every $c \in (a, b)$, there exists some $x_0 \in G$ such that $f(x_0) = c$.

Let us now look at other types of continuity for functions.

It is known that for a given function $f(\cdot)$ and $\bar{x}$, the limit $\lim\limits_{x \to \bar{x}} f(x)$ does not necessarily exist. Therefore, we would like to look at a little weaker notions. For any bounded function $f : G \to \mathbb{R}$, and $\bar{x} \in \bar{G}$, define

$$\begin{aligned}
f^-(\bar{x}\,;\delta) &= \inf f\big(G \cap B_\delta(\bar{x}) \setminus \{\bar{x}\}\big), \\
f^+(\bar{x}\,;\delta) &= \sup f\big(G \cap B_\delta(\bar{x}) \setminus \{\bar{x}\}\big).
\end{aligned} \tag{3.4}$$

It is easy to see that $\delta \mapsto f^-(\bar{x}\,;\delta)$ is non-increasing and $\delta \mapsto f^+(\bar{x}\,;\delta)$ is non-decreasing. Hence, the limits on the right-hand side of the following exist:

$$\varliminf_{x \to \bar{x}} f(x) = \lim_{\delta \downarrow 0} f^-(\bar{x}\,;\delta), \qquad \varlimsup_{x \to \bar{x}} f(x) = \lim_{\delta \downarrow 0} f^+(\bar{x}\,;\delta). \qquad (3.5)$$

They are called *limit inferior* and *limit superior* (or simply called liminf and limsup), respectively, of $f(\cdot)$ as x approaches to $\bar{x}$. Note that for any bounded function $f(\cdot)$, liminf and limsup always exist. Here is a very typical example. Let

$$f(x) = \sin\frac{1}{x}, \qquad x \neq 0.$$

Then it is straightforward that

$$\varliminf_{x \to 0} f(x) = -1, \qquad \varlimsup_{x \to 0} f(x) = 1.$$

We have the following result which is comparable with Proposition 1.16.

Proposition 3.4. *Let $G \subseteq \mathbb{R}^n$ be a domain and $x_0 \in \bar{G}$. Let $f : G \to \mathbb{R}$. Then*

$$\varliminf_{x \to x_0} f(x) \leqslant \varlimsup_{x \to x_0} f(x). \qquad (3.6)$$

Further,

$$\lim_{x \to x_0} f(x) \text{ exists} \quad \Longleftrightarrow \quad \varliminf_{x \to x_0} f(x) = \varlimsup_{x \to x_0} f(x). \qquad (3.7)$$

In this case,

$$\lim_{x \to x_0} f(x) = \varliminf_{x \to x_0} f(x) = \varlimsup_{x \to x_0} f(x). \qquad (3.8)$$

Next, we introduce the following notions which are generalizations of the continuity.

Definition 3.5. *Let $f : G \to \mathbb{R}$ and $\bar{x} \in G$.*

(i) *$f(\cdot)$ is lower semi-continuous at $\bar{x}$ if*

$$f(\bar{x}) \leqslant \varliminf_{x \to \bar{x}} f(x).$$

(ii) *$f(\cdot)$ is upper semi-continuous at $\bar{x}$ if*

$$f(\bar{x}) \geqslant \varlimsup_{x \to \bar{x}} f(x).$$

The following examples are worthy of remembering: Let

$$\varphi_-(x) = \begin{cases} -1, & x = 0, \\ 0, & x \neq 0, \end{cases} \qquad \varphi_+(x) = \begin{cases} 1, & x = 0, \\ 0, & x \neq 0. \end{cases} \qquad (3.9)$$

Then $\varphi_-(\cdot)$ is lower semi-continuous and $\varphi_+(\cdot)$ is upper semi-continuous.

It is clear that $f(\cdot)$ is lower semi-continuous if and only if $-f(\cdot)$ is upper semi-continuous. The following result reveals some further interesting relations among lower semi-continuity, upper semi-continuity and continuity of functions.

Theorem 3.6. *Let $G \subseteq \mathbb{R}^n$ and $f : G \to \mathbb{R}$.*

(i) *$f(\cdot)$ is lower semi-continuous on G if and only if for every $a \in \mathbb{R}$,*

$$\{f > a\} \triangleq \{x \in G \mid f(x) > a\}$$

is open in G.

(ii) *$f(\cdot)$ is upper semi-continuous on G if and only if for every $a \in \mathbb{R}$,*

$$\{f < a\} \triangleq \{x \in G \mid f(x) < a\}$$

is open in G.

(iii) *$f(\cdot)$ is continuous on G if and only if for every $a \in \mathbb{R}$, both $\{f < a\}$ and $\{f > a\}$ are open in G.*

(iv) *$f(\cdot)$ is continuous on G if and only if for every open set $V \subseteq \mathbb{R}$, the set*

$$f^{-1}(V) \triangleq \{x \in G \mid f(x) \in V\}$$

is open in G, i.e., there exists an open set $U \subseteq \mathbb{R}^n$ such that

$$f^{-1}(V) = G \cap U.$$

The proof of the above result follows from the definitions of the continuity, the lower and upper semi-continuity. Let us present a case for which the above result plays an interesting role.

Proposition 3.7. *Let $G \subseteq \mathbb{R}^n$ be a compact set and let $f_\lambda : G \to \mathbb{R}$ be a family of functions, with $\lambda \in \Lambda$, for some index set Λ.*

(i) *Suppose each $f_\lambda(\cdot)$ is upper semi-continuous on G, so is $f_+(\cdot)$ where*

$$f_+(x) \triangleq \inf_{\lambda \in \Lambda} f_\lambda(x), \qquad \forall x \in G.$$

(ii) *Suppose each $f_\lambda(\cdot)$ is lower semi-continuous on G, so is $f_-(\cdot)$ where*

$$f_-(x) \triangleq \sup_{\lambda \in \Lambda} f_\lambda(x), \qquad \forall x \in G.$$

(iii) *In particular, if each $f_\lambda(\cdot)$ is continuous on G, then $f_+(\cdot)$ is upper semi-continuous and $f_+(\cdot)$ is lower semi-continuous.*

Note that in the above part (iii), functions $f_\pm(\cdot)$ are not necessarily continuous. Let us look at the following examples.

Example 3.8. Let

$$f_k(x) = (1 - |x|)^k, \qquad x \in [-1, 1], \ k \geqslant 1.$$

Then each $f_k(\cdot)$ is continuous and is non-increasing in k. Hence,

$$\inf_{k \geqslant 1} f_k(x) = \lim_{k \to \infty} f_k(x) = \varphi_+(x), \qquad x \in [-1, 1],$$

where $\varphi_+(\cdot)$ is defined in (3.9), which is upper semi-continuous, and is not continuous. Similarly, if we define

$$g_k(x) = -(1 - |x|)^k, \qquad x \in [-1, 1], \ k \geqslant 1,$$

then

$$\sup_{k \geqslant 1} g_k(x) = \lim_{k \to \infty} g_k(x) = \varphi_-(x), \qquad x \in [-1, 1],$$

where $\varphi_-(\cdot)$ is defined in (3.9), which is lower semi-continuous, and is not continuous.

Example 3.9. Let $F \subseteq \mathbb{R}^n$ be a closed set and define the *indicator function* of F by the following:

$$I_F(x) = \begin{cases} 1, & x \in F, \\ 0, & x \in F^c. \end{cases}$$

Then $I_F(\cdot)$ is upper semi-continuous, and $I_{F^c}(\cdot) = 1 - I_F(\cdot)$ is lower semi-continuous.

Let us now briefly look at the case of *vector-valued* functions, by which we mean maps $f : G \to \mathbb{R}^m$ with $m \geqslant 2$. In this case, we may write

$$f(x) = (f_1(x), \cdots, f_m(x))^\top, \qquad x \in G,$$

where each $f_i : G \to \mathbb{R}$ is a scalar function (of multi-variables). For such kind of functions, we can define limit and continuity the same as scalar functions (see Definition 3.1) with (3.1) replaced by

$$\|f(x) - L\| < \varepsilon,$$

and $L = (L_1, \cdots, L_m)^\top \in \mathbb{R}^m$. Note that for vector-valued functions, we do not define limsup and liminf, as well as upper semi-continuity and lower semi-continuity. The following result makes the study of vector-valued functions very simple.

Proposition 3.10. *Let* $f(\cdot) = (f_1(\cdot), \cdots, f_m(\cdot))^\top : G \to \mathbb{R}^m$ *be a vector-valued function.*

(i) $\lim\limits_{x \to \bar{x}} f(x)$ *exists if and only if* $\lim\limits_{x \to \bar{x}} f_i(x)$ *exists for each* $i = 1, \cdots, m$;

(ii) $f(\cdot)$ *is continuous at* $\bar{x}$ *if and only if so is each* $f_i(\cdot)$, $1 \leqslant i \leqslant m$.

Because of the above proposition, many results for scalar-valued functions have their vector-valued function counterparts. We omit the details here.

Exercises

1. Let $f, g : G \to \mathbb{R}$ be continuous on G. Then $f \pm g$, fg, $f \vee g$ and $f \wedge g$ are also continuous on G, where $f \vee g = \max\{f, g\}$ and $f \wedge g = \min\{f, g\}$. What happens if f and g are upper semi-continuous? And how about if f and g are lower semi-continuous?

2. Let $f_k(x) = \sin^k x$, $x \in [0, \pi]$, $k \geqslant 1$. Find

$$F^+(x) = \sup_{k \geqslant 1} f_k(x), \quad F^-(x) = \inf_{k \geqslant 1} f_k(x), \qquad x \in [0, \pi].$$

Are $F^\pm(x)$ continuous? Give your reasons.

3. Find the following:

(i) $\varlimsup\limits_{x \to \infty} \sin x$; (ii) $\varliminf\limits_{x \to \infty} \cos x$; $\varlimsup\limits_{x \to 0} \dfrac{\sin x}{x}$; $\varliminf\limits_{x \to 0} \dfrac{\tan x}{x}$.

4. Let $G \subseteq \mathbb{R}^n$ and $f : G \to \mathbb{R}^m$. Define

$$f^{-1}(B) = \{x \in G \mid f(x) \in B\}, \qquad \forall B \subseteq \mathbb{R}^m.$$

Let $A_\lambda \subseteq \mathbb{R}^m$ be a family of sets with $\lambda \in \Lambda$. Show that

$$f^{-1}\left(\bigcup_{\lambda \in \Lambda} A_\lambda\right) = \bigcup_{\lambda \in \Lambda} f^{-1}(A_\lambda), \quad f^{-1}\left(\bigcap_{\lambda \in \Lambda} A_\lambda\right) = \bigcap_{\lambda \in \Lambda} f^{-1}(A_\lambda).$$

5. Prove Theorem 3.6.

6*. Let $G \subseteq \mathbb{R}^n$ be compact and $f_k : G \to \mathbb{R}$ be a family of functions satisfying

$$|f_k(x) - f_k(y)| \leqslant L\|x - y\|, \qquad \forall x, y \in G, \ k \geqslant 1.$$

Let $F^{\pm}(\cdot)$ be defined by

$$F^+(x) = \sup_{k \geqslant 1} f_k(x), \quad F^-(x) = \inf_{k \geqslant 1} f_k(x), \qquad x \in G.$$

Show that

$$|F^{\pm}(x) - F^{\pm}(y)| \leqslant L\|x - y\|, \qquad \forall x, y \in G.$$

7. Let $A \in \mathbb{R}^{n \times n}$ with $\|A\| < 1$. Find $\lim_{k \to \infty} \|A^k\|$.

8*. Prove that if $A \in \mathbb{S}^n$ is positive definite, then there exists a $\delta > 0$ such that

$$x^{\top} A x \geqslant \delta\|x\|_2^2, \qquad \forall x \in \mathbb{R}^n.$$

4   Differentiation

Let us first recall the situation of single-variable functions.

Definition 4.1. Let $f : [a, b] \to \mathbb{R}$ and $x_0 \in (a, b)$. We say that $f(\cdot)$ admits a *derivative* $f'(x_0)$ at x_0 if the following limit exists:

$$f'(x_0) \equiv \lim_{\delta \to 0} \frac{f(x_0 + \delta) - f(x_0)}{\delta}. \tag{4.1}$$

In this case, we also say that $f(\cdot)$ is *differentiable* at x_0. If $f(\cdot)$ is differentiable at every point $x \in (a, b)$, we say that $f(\cdot)$ is differentiable on (a, b). Further, if $f(\cdot)$ is continuous on $[a, b]$ and differentiable on (a, b), with $f'(\cdot)$ being continuous on (a, b) and can be extended to a continuous function on $[a, b]$, i.e., there exists a continuous function $g : [a, b] \to \mathbb{R}$ such that

$$f'(x) = g(x), \qquad \forall x \in (a, b),$$

then we say that $f(\cdot)$ is C^1 on $[a, b]$. In a similar fashion, we can define $f(\cdot)$ to be C^k, for any $k \geqslant 1$.

For computational aspect of derivatives, there are some important rules, such as linearity rule, product rule, quotient rule, chain rule, etc. All these make the calculation of derivatives efficient and one does not have to use the definition every time when a derivative of certain function needs to be calculated. We leave the details to the readers. The following result is simple but very deep.

Theorem 4.2. (Fermat) *Let $f : (a, b) \to \mathbb{R}$. If $f(\cdot)$ attained a local extreme at $x_0 \in (a, b)$ and $f'(x_0)$ exists, then*

$$f'(x_0) = 0. \tag{4.2}$$

The proof of the above result follows easily from the definitions of local extreme and the derivative.

Theorem 4.3. (Rolle) *Let $f : [a, b] \to \mathbb{R}$ be continuous and admits a derivative at any point on (a, b). Let*

$$f(a) = f(b) = 0. \tag{4.3}$$

Then there exists a $c \in (a, b)$ such that

$$f'(c) = 0. \tag{4.4}$$

Proof. If $f(x) \equiv 0$, then by definition of derivative, we see that (4.4) must be true for every $c \in (a, b)$. If $f(x) \not\equiv 0$, then by Theorem 3.3, we know that $f(\cdot)$ attains a global extreme (either minimum or maximum) over $[a, b]$ at some point $c \in (a, b)$, which is different from a and b. Then Theorem 4.2 applies. $\qquad\square$

Theorem 4.4. (Cauchy Mean-Value Theorem) *Let $f, g : [a, b] \to \mathbb{R}$ be C^1. Then there exists a $\xi \in (a, b)$ such that*

$$g'(\xi)[f(b) - f(a)] = f'(\xi)[g(b) - g(a)]. \tag{4.5}$$

In particular, by taking $g(x) = x$, one has

$$f(b) - f(a) = f'(\xi)(b - a). \tag{4.6}$$

Proof. First, if $g(a) = g(b)$, we let

$$h(t) = g(a + t(b - a)) - g(a), \qquad t \in [0, 1].$$

Then $h : [0, 1] \to \mathbb{R}$ is continuous and differentiable. Moreover, $h(0) = h(1) = 0$. Thus, by Rolle's theorem, there exists a $\theta \in (0, 1)$ such that (by means of chain rule)

$$0 = h'(\theta) = g'(a + \theta(b - a))(b - a).$$

Hence, taking $\xi = a + \theta(b - a) \in (a, b)$, we see that both sides of (4.5) are zero.

Next, we assume that $g(a) \neq g(b)$. Let

$$h(t) = f(a + t(b - a)) - f(a)$$
$$- \frac{f(b) - f(a)}{g(b) - g(a)} [g(a + t(b - a)) - g(a)], \qquad t \in [0, 1].$$

Then $h : [0, 1] \to \mathbb{R}$ is continuous and admits a derivative on $(0, 1)$. Moreover,

$$h(0) = h(1) = 0.$$

Thus, by the Rolle's theorem, there exists a $\theta \in (0,1)$ such that

$$0 = h'(\theta) = f'(a + \theta(b-a))(b-a)$$
$$- \frac{f(b)-f(a)}{g(b)-g(a)} g'(a + \theta(b-a))(b-a).$$

If we call $\xi = a + \theta(b-a) \in (a,b)$, (4.5) follows. $\qquad\square$

Theorem 4.5. (Taylor) *Let* $f : [a,b] \to \mathbb{R}$ *be* C^{k+1} *and* $x_0 \in (a,b)$. *Then for any* $x \in [a,b]$, *there exists a* ξ *between* x_0 *and* x *such that*

$$f(x) = f(x_0) + \frac{f'(x_0)}{1!}(x-x_0) + \frac{f''(x_0)}{2!}(x-x_0)^2 + \cdots \tag{4.7}$$
$$+ \frac{f^{(k)}(x_0)}{k!}(x-x_0)^k + \frac{f^{(k+1)}(\xi)}{(k+1)!}(x-x_0)^{k+1}.$$

Proof. Fix an $x \in (a,b)$. Let

$$\begin{cases} \bar{f}(t) = f(t) + \dfrac{f'(t)}{1!}(x-t) + \dfrac{f''(t)}{2!}(x-t)^2 + \cdots + \dfrac{f^{(k)}(t)}{k!}(x-t)^k, \\ \bar{g}(t) = (x-t)^{k+1}. \end{cases}$$

Then by Cauchy Mean-Value Theorem, we have some ξ between x_0 and x such that

$$\frac{\bar{f}'(\xi)}{\bar{g}'(\xi)} = \frac{\bar{f}(x) - \bar{f}(x_0)}{\bar{g}(x) - \bar{g}(x_0)}.$$

Note that

$$\bar{f}'(t) = f'(t) - f'(t) + \frac{f''(t)}{1!}(x-t) - \frac{f''(t)}{1!}(x-t) + \cdots$$
$$+ \frac{f^{(k)}(t)}{(k-1)!}(x-t)^{k-1} - \frac{f^{(k)}(t)}{(k-1)!}(x-t)^{k-1} + \frac{f^{(k+1)}(t)}{k!}(x-t)^k$$
$$= \frac{f^{(k+1)}(t)}{k!}(x-t)^k,$$

and

$$\bar{g}'(t) = -(k+1)(x-t)^k.$$

Thus,

$$-\frac{f^{(k+1)}(\xi)}{(k+1)!} = \frac{\bar{f}'(\xi)}{\bar{g}'(\xi)} = \frac{\bar{f}(x) - \bar{f}(x_0)}{\bar{g}(x) - \bar{g}(x_0)} = \frac{f(x) - \bar{f}(x_0)}{-\bar{g}(x_0)}$$
$$= \frac{1}{-(x-x_0)^{k+1}} \left[f(x) - \left(f(x_0) + \frac{f'(x_0)}{1!}(x-x_0) \right.\right.$$
$$\left.\left. + \frac{f''(x_0)}{2!}(x-x_0)^2 + \cdots + \frac{f^{k)}(x_0)}{k!}(x-x_0)^k \right) \right].$$

Hence, (4.5) follows. □

Now, we turn to multi-variable functions.

Definition 4.6. Let $G \subseteq \mathbb{R}^n$ be a domain, $\bar{x} \in G$, and $f : G \to \mathbb{R}$ be a function.

(i) $f(\cdot)$ is *directionally differentiable* at $\bar{x}$ in the direction $u \in \partial B_1(0)$ if the following limit exists:

$$f'(\bar{x}; u) = \lim_{\delta \downarrow 0} \frac{f(\bar{x} + \delta u) - f(\bar{x})}{\delta},$$

which is called the *directional derivative* of $f(\cdot)$ at $\bar{x}$ in the direction u.

(ii) $f(\cdot)$ is *Gâteaux differentiable* at $\bar{x}$ if it is directionally differentiable in every direction $u \in \partial B_1(0)$, i.e., $u \mapsto f'(\bar{x}; u)$ is a well-defined map (on $\partial B_1(0)$) which is called the *Gâteaux derivative* of $f(\cdot)$ at $\bar{x}$.

(iii) $f(\cdot)$ admits a *gradient*, denoted by $\nabla f(\bar{x}) \in \mathbb{R}^{1 \times n}$, at $\bar{x}$ if $f(\cdot)$ is Gâteaux differentiable at $\bar{x}$ and the Gâteaux derivative $f'(\bar{x}; \cdot)$ admits the following (linear) representation:

$$f'(\bar{x}; u) = [\nabla f(\bar{x})]u, \qquad \forall u \in \partial B_1(0).$$

(iv) $f(\cdot)$ is *Fréchet differentiable* at $\bar{x}$ if there exists an $f_x(\bar{x}) \in \mathbb{R}^{1 \times n}$, called the *Fréchet derivative* of $f(\cdot)$ at $\bar{x}$, such that

$$\lim_{G \ni x \to \bar{x}} \frac{|f(x) - f(\bar{x}) - f_x(\bar{x})(x - \bar{x})|}{\|x - \bar{x}\|} = 0. \tag{4.8}$$

From (4.8), we see that if $f : G \to \mathbb{R}$ is Fréchet differentiable at $\bar{x}$, then

$$\begin{cases} f(x) = f(\bar{x}) + f_x(\bar{x})(x - \bar{x}) + R_1(\bar{x}; x), & \forall x \in G, \\ \displaystyle \lim_{G \ni x \to \bar{x}} \frac{R_1(\bar{x}; x)}{\|x - \bar{x}\|} = 0. \end{cases} \tag{4.9}$$

We refer to (4.9) as the *first order Taylor expansion* of (multi-variable function) $f(\cdot)$ at $\bar{x}$. From such an expansion, we see immediately that the Fréchet differentiability implies the continuity.

In the rest of this section, we let $G \subseteq \mathbb{R}^n$ be a domain. The following proposition collects some basic results concerning the above notions.

Proposition 4.7. (i) *If $f : G \to \mathbb{R}$ is Gâteaux differentiable at $\bar{x}$, then for every $u \in \mathbb{R}^n$, the following is well-defined.*

$$f'(\bar{x}; u) \triangleq \lim_{\delta \downarrow 0} \frac{f(\bar{x} + \delta u) - f(\bar{x})}{\delta} = \begin{cases} 0, & u = 0, \\ \|u\| f'\left(\bar{x}; \dfrac{u}{\|u\|}\right), & u \neq 0. \end{cases}$$

Thus, in this case, $u \mapsto f'(\bar{x}; u)$ is defined on $\mathbb{R}^n$ and it is positively homogeneous, i.e.,

$$f'(\bar{x}; \lambda u) = \lambda f'(\bar{x}; u), \qquad \forall \lambda \in [0, \infty), \ u \in \mathbb{R}^n.$$

(ii) *If $f : G \to \mathbb{R}$ is Fréchet differentiable at $\bar{x}$, then it admits a gradient $\nabla f(\bar{x})$ at $\bar{x}$, and*

$$\nabla f(\bar{x}) = f_x(\bar{x}).$$

Therefore, the following implications hold:

$$
\begin{aligned}
\text{Fréchet differentiability} &\Rightarrow \text{existence of gradient} \\
&\Rightarrow \text{Gâteaux differentiability.}
\end{aligned}
\tag{4.10}
$$

(iii) *If $f : G \to \mathbb{R}$ is Fréchet differentiable at $\bar{x}$, then it is continuous at $\bar{x}$.*

Note that the reverse implications in (4.10) are not true in general. The main points here are the following:

• The Gâteaux derivative $u \mapsto f'(\bar{x}; u)$ is not necessarily additive in u, i.e.,

$$f'(\bar{x}; u + v) = f'(\bar{x}; u) + f'(\bar{x}; v)$$

might fail in general. Therefore, Gâteaux differentiability does not imply the existence of a gradient.

• The existence of a gradient does not imply the continuity, whereas the Fréchet differentiability implies the continuity. Hence, the existence of a gradient does not imply the Fréchet differentiability.

Example 4.8. Let $f(x) = |x|$ for $x \in \mathbb{R}$. For $u = \pm 1$,

$$f'(0; u) = \lim_{\delta \downarrow 0} \frac{f(\delta u) - f(0)}{\delta} = \lim_{\delta \to 0} \frac{|\delta u|}{\delta} = |u| = 1.$$

Clearly, for this $f(\cdot)$, $\nabla f(0)$ does not exists. Hence, this is an example that a Gâteaux differentiable function does not admit a gradient.

Note that in the above, the various differentiability are defined without using coordinates. Now, if $f(\cdot)$ is Gâteaux differentiable and for some $1 \leqslant i \leqslant n$,

$$f'(\bar{x}; -e_i) = -f'(\bar{x}; e_i),$$

where $e_i \in \mathbb{R}^n$ is the vector whose i-th component is 1 and all other components are 0, then we define

$$f_{x_i}(\bar{x}) = f'(\bar{x}; e_i),$$

and call it the *partial derivative* of $f(\cdot)$ with respect to x_i. We have the following result.

Proposition 4.9. *Let $f : G \to \mathbb{R}$.*

(i) *Suppose that $f(\cdot)$ admits a gradient $\nabla f(\bar{x})$ at $\bar{x} \in G$, then $f_{x_i}(\bar{x})$ exists for all $i = 1, \cdots, n$, and in this case, the gradient $\nabla f(\bar{x})$ admits the following representation:*

$$\nabla f(\bar{x}) = (f_{x_1}(\bar{x}), \cdots, f_{x_n}(\bar{x})) \in \mathbb{R}^{1 \times n}.$$

Consequently,

$$f_{x_i}(\bar{x}) = \nabla f(\bar{x}) e_i, \qquad 1 \leq i \leq n.$$

(ii) *Suppose for each $i = 1, \cdots, n$, $f_{x_i}(\cdot)$ is continuous in a neighborhood of $\bar{x}$, then $f(\cdot)$ is Fréchet differentiable at $\bar{x}$.*

Hereafter, when $f(\cdot)$ is Fréchet differentiable at every point in G and $f_x(\cdot) e_i \equiv f_{x_i}(\cdot)$ is continuous on G for each $i = 1, \cdots, n$, we say that $f(\cdot)$ is C^1 in G.

Theorem 4.10. (Mean Value Theorem) *Let $f : G \to \mathbb{R}$ be C^1 on G and $x, y \in G$ such that*

$$[x, y] \triangleq \{(1 - \lambda)x + \lambda y \mid \lambda \in [0, 1]\} \subseteq G.$$

Then there exists a $\theta \in (0, 1)$ such that

$$f(y) - f(x) = f_x(x + \theta(y - x))(y - x). \tag{4.11}$$

Proof. Let

$$h(t) = f(x + t(y - x)), \qquad t \in [0, 1].$$

Then by Theorem 4.4, we have

$$f(y) - f(x) = h(1) - h(0) = h'(\theta) = f_x(x + \theta(y - x))(y - x),$$

for some $\theta \in (0, 1)$. $\qquad\qquad\qquad\qquad\qquad\qquad\qquad\qquad\qquad \square$

Note that replacing x by $\bar{x}$ and y by x, (4.11) is equivalent to the following:

$$f(x) = f(\bar{x}) + f_x(\bar{x} + \theta(x - \bar{x}))(x - \bar{x}). \tag{4.12}$$

We then can make a comparison between the above with (4.9): the trade-off of having no remainder term $R_1(\bar{x}; x)$ is to replace $\bar{x}$ by $\bar{x} + \theta(x - \bar{x})$ in $f_x(\cdot)$ above.

Next, let $f : G \to \mathbb{R}$ be C^1. Further, suppose each $f_x(\cdot)e_i \equiv f_{x_i}(\cdot)$ also admits a gradient. Then we can define the *mixed partial derivative*:

$$f_{x_i x_j}(\bar{x}) \triangleq (f_{x_i})_{x_j}(\bar{x}) \equiv [\nabla f_{x_i}(\bar{x})]e_j. \tag{4.13}$$

It is natural to ask if $f_{x_i x_j}(\cdot)$ is the same as $f_{x_j x_i}(\cdot)$. In general, this is not true. However, we have the following result.

Theorem 4.11. (Clairaut's Theorem) *Suppose for given* $1 \leqslant i, j \leqslant n$, $f_{x_i x_j}(\cdot)$ *exist in a neighborhood of* $\bar{x}$ *and are all continuous at* $\bar{x}$. *Then*

$$f_{x_i x_j}(\bar{x}) = f_{x_j x_i}(\bar{x}), \qquad 1 \leqslant i, j \leqslant n.$$

When the above happens, we let

$$f_{xx}(x) = (f_{x_i x_j}(x)), \qquad \forall x \in G,$$

and call it the *Hessian matrix* of $f(\cdot)$, which is symmetric. Next result will be very useful later.

Theorem 4.12. *Let* $f : G \to \mathbb{R}$ *such that each* $f_{x_i}(\cdot)$ *is* C^1 *in* G $(1 \leqslant i \leqslant n)$. *Let* $x, \bar{x} \in G$ *such that*

$$[\bar{x}, x] \equiv \big\{ (1 - \lambda)\bar{x} + \lambda x \mid \lambda \in [0, 1] \big\} \subseteq G^\circ.$$

Then there exists a $\xi \in [\bar{x}, x]$ *such that*

$$f(x) - f(\bar{x}) = f_x(\bar{x})(x - \bar{x}) + \frac{1}{2}(x - \bar{x})^\top f_{xx}(\xi)(x - \bar{x}). \tag{4.14}$$

Proof. Let

$$h(t) = f(\bar{x} + t(x - \bar{x})), \qquad t \in [-\delta, 1 + \delta],$$

for some $\delta > 0$. By Theorem 4.5, we have

$$f(x) = h(1) = h(0) + h'(0) + \frac{1}{2}h''(\theta)$$
$$= f(\bar{x}) + f_x(\bar{x})(x - \bar{x}) + \frac{1}{2}(x - \bar{x})^\top f_{xx}(\bar{x} + \theta(x - \bar{x}))(x - \bar{x}),$$

for some $\theta \in (0, 1)$. Hence, (4.14) follows by letting $\xi = \bar{x} + \theta(x - \bar{x})$. $\quad\square$

We refer to (4.14) as the *second order Taylor expansion* of $f(\cdot)$ at $\bar{x}$.

Next, we consider vector-valued function $f : G \to \mathbb{R}^m$. We may define all kinds of differentiabilities of such an $f(\cdot)$ in a way similar to Definition 4.6, with now $f'(\bar{x}; u) \in \mathbb{R}^m$, and

$$\nabla f(\bar{x}) = \begin{pmatrix} \nabla f_1(\bar{x}) \\ \vdots \\ \nabla f_m(\bar{x}) \end{pmatrix}, \quad f_x(\bar{x}) = \begin{pmatrix} (f_1)_x(\bar{x}) \\ \vdots \\ (f_m)_x(\bar{x}) \end{pmatrix} \in \mathbb{R}^{m \times n}.$$

On the other hand, we have the following result.

Proposition 4.13. *Let* $f(\cdot) \equiv (f_1(\cdot), \cdots, f_m(\cdot))^\top : G \to \mathbb{R}^m$. *Then* $f(\cdot)$ *is Gâteaux differentiable (admits a gradient, is Fréchet differentiable, is* C^1, *is* C^2, *etc.), so is (does) each component* $f_i(\cdot)$, $1 \leqslant i \leqslant m$.

Now, we consider a map $f : \mathbb{R}^n \times \mathbb{R}^m \to \mathbb{R}^m$. Let the vectors in $\mathbb{R}^n$ and $\mathbb{R}^m$ be denoted by x and z, respectively. Then for fixed $x \in \mathbb{R}^n$, we may talk about the differentiability of $z \mapsto f(x, z)$, and for fixed $z \in \mathbb{R}^m$, we may talk about the differentiability of $x \mapsto f(x, z)$. They are kind of partial derivatives. If $f(\cdot, \cdot)$ is C^1, then $f_x(x, z) \in \mathbb{R}^{m \times n}$ and $f_z(x, z) \in \mathbb{R}^{m \times m}$, for each fixed (x, z). We now state the following important result which will be used in a late chapter.

Theorem 4.14. (Implicit Function Theorem) *Let* $G \subseteq \mathbb{R}^n \times \mathbb{R}^m$ *be a domain and* $f : G \to \mathbb{R}^m$ *be* C^1. *Let* $(\bar{x}, \bar{z}) \in G$ *such that*

$$f(\bar{x}, \bar{z}) = 0,$$

and

$$\det\left(f_z(\bar{x}, \bar{z})\right) \neq 0.$$

Then there exists a neighborhood V *of* $\bar{x}$ *and a* C^1 *map* $\varphi : V \to \mathbb{R}^m$ *such that*

$$f(x, \varphi(x)) = 0, \qquad \forall x \in V, \qquad \varphi(\bar{x}) = \bar{z},$$

and

$$\varphi_x(x) = -f_z(x, \varphi(x))^{-1} f_x(x, \varphi(x)), \qquad x \in V.$$

We call $\varphi(\cdot)$ in the above an *implicit function* determined by the equation $f(x, z) = 0$. We omit the proof here.

Exercises

1. Calculate $f_x(x)$ and $f_{xx}(x)$ for the following $f(x)$:

(i) $f(x) = (a^\top x)(b^\top x)$, with $a, b \in \mathbb{R}^n$.

(ii) $f(x) = x^\top A x$, with $A \in \mathbb{S}^n$. What happens if $A \in \mathbb{R}^{n \times n}$ is not symmetric?

(iii) $f(x) = \|Ax - b\|^2$, with $A \in \mathbb{R}^{m \times n}$ and $b \in \mathbb{R}^m$.

(iv) $f(x) = \sin(a^\top x)$, with $a \in \mathbb{R}^n$.

2. Find the second order Taylor expansion of $f(x)$ at the given point x_0:

(i) $f(x) = (x_1 - x_2)^2 + x_2 + 1$; $x_0 = (1, 0)^\top$.

(ii) $f(x) = (x-a)^\top A(x-a) + 2b^\top(x-a) + c$ with some $A \in \mathbb{S}^n$, $a, b \in \mathbb{R}^n$ and $c \in \mathbb{R}$; $x_0 \in \mathbb{R}^n$.

3. Let $f(x, y)$ be given by the following:
$$f(x, y) = \begin{cases} \dfrac{xy}{\sqrt{x^2 + y^2}}, & (x, y) \neq (0, 0), \\ 0, & (x, y) = (0, 0). \end{cases}$$
Show that $f(\cdot, \cdot)$ is Gâteaux differentiable at $(0, 0)$, but it does not admit a gradient at $(0, 0)$.

4. Let $f(x, y)$ be given by the following:
$$f(x, y) = \begin{cases} \dfrac{x^3 y}{x^6 + y^2}, & (x, y) \neq (0, 0), \\ 0, & (x, y) = (0, 0). \end{cases}$$
Show that $f(\cdot, \cdot)$ admits a gradient at $(0, 0)$, but it is not Fréchet differentiable at $(0, 0)$. (Hint: Show that this function is discontinuous at $(0, 0)$.)

5. Let $f(x, y)$ be given by the following:
$$f(x, y) = \begin{cases} xy\left(\dfrac{x^2 - y^2}{x^2 + y^2}\right), & (x, y) \neq (0, 0), \\ 0, & (x, y) = (0, 0). \end{cases}$$
Calculate $f_{xy}(0, 0)$ and $f_{yx}(0, 0)$. Comment on your result.

6. Let $f(x) = \|x\|$ for $x \in \mathbb{R}^n$. Is $f(\cdot)$ Gâteaux differentiable? Fréchet differentiable? Why?

7*. Show that there exist a C^1 function $\varphi : V \to \mathbb{R}$ with V being a neighborhood of $(1, 1)$ such that $\varphi(1, 1) = 0$ and
$$\varphi(x, y) \sin[\varphi(x, y)] + x^2 e^y + xy\varphi(x, y) = 0.$$
Find $\varphi_x(x, y)$ and $\varphi_y(x, y)$, for $(x, y) \in V$.

Chapter 2

Optimization Problems and Existence of Optimal Solutions

1 Optimization Problems

Let $F, G \subseteq \mathbb{R}^n$ be non-empty sets with $F \subseteq G$. Let $f : G \to \mathbb{R}$ be a given function. It is possible that $G = \mathbb{R}^n$ and/or $G = F$. We pose the following *optimization problem*:

Problem (G). Find an $\bar{x} \in F$ such that

$$f(\bar{x}) = \min_{x \in F} f(x). \tag{1.1}$$

In the above, $f(\cdot)$ is called an *objective function*, F is called a *feasible set* (or a *constraint set*). Any $x \in F$ is called a *feasible point* (or a *feasible solution*) of Problem (G), and any $\bar{x} \in F$ satisfying (1.1) is called an *optimal solution* of Problem (G). Since this problem is to find a minimum of the function $f(\cdot)$ over the set F, we also refer to it as a *minimization problem*. As Definition 3.2 of Chapter 1, any $\bar{x} \in F$ satisfying (1.1) is called a *global minimum* of $f(\cdot)$ over F. Correspondingly, we say that $\bar{x} \in F$ is a *local minimum* of $f(\cdot)$ over F if there exists a $\delta > 0$ such that $\bar{x}$ is a global minimum of $f(\cdot)$ over $F \cap B_\delta(\bar{x})$.

We may likewise pose the following optimization problem.

Problem (G′). Find an $\bar{x} \in F$ such that

$$f(\bar{x}) = \max_{x \in F} f(x). \tag{1.2}$$

This problem is referred to as a *maximization problem*. Similar to the above, we may define *global maximum* and *local maximum*. Sometimes, optimization problems are also called *mathematical programming* problems. It is clear that Problem (G′) is equivalent to Problem (G) with $f(\cdot)$ replaced by $-f(\cdot)$. Thus, it suffices to concentrate on one of them, say, Problem (G).

The goal of optimization problem is to find optimal solutions, and study properties of optimal solutions. Therefore, the following topics are basic:

(i) Existence of optimal solutions;

(ii) Characterization of optimal solutions;

(iii) Computation algorithms for finding approximate optimal solutions.

These problems will be discussed in some extent in the following chapters.

Some special cases of Problem (G) are of particular interest. Let us list some of them. First, let $f : G \to \mathbb{R}$, $g : G \to \mathbb{R}^m$ and $h : G \to \mathbb{R}^\ell$, with $G \subseteq \mathbb{R}^n$,

$$g(x) = \begin{pmatrix} g^1(x) \\ g^2(x) \\ \vdots \\ g^m(x) \end{pmatrix}, \qquad h(x) = \begin{pmatrix} h^1(x) \\ h^2(x) \\ \vdots \\ h^\ell(x) \end{pmatrix}, \qquad x \in \mathbb{R}^n.$$

Consider the following problem:

$$\begin{cases} \min \ f(x), \\ \text{subject to } g^i(x) = 0, & 1 \leqslant i \leqslant m, \\ \qquad\qquad\quad h^j(x) \leqslant 0, & 1 \leqslant j \leqslant \ell. \end{cases} \tag{1.3}$$

Each $g^i(x) = 0$ is called an *equality constraint* and each $h^j(x) \leqslant 0$ is called an *inequality constraint*. Thus, in the above, there are m equality constraints and ℓ inequality constraints. We write the above more compactly as follows:

$$\begin{cases} \min \ f(x), \\ \text{subject to } g(x) = 0, \ h(x) \leqslant 0, \end{cases} \tag{1.4}$$

where $h(x) \leqslant 0$ stands for componentwise inequalities. We see that the above (1.4) is a special case of Problem (G) with

$$F = \{x \in G \mid g(x) = 0, \ h(x) \leqslant 0\}. \tag{1.5}$$

The above (1.4) is usually referred to as a *nonlinear programming problem* (NLP problem, for short). For such a problem, we at least should assume the following:

$$n > m. \tag{1.6}$$

Namely, there are more *decision variables* $x_1, \cdots, x_n$ than the number of equality constraints. Otherwise, it is quite possible that either $F = \varnothing$ or F is a singleton, which makes the corresponding optimization problem trivial/meaningless.

A further special case is the so-called *linear programming* problem (LP problem, for short). A typical form of such a problem looks like the following:

$$\begin{cases} \min \ c^\top x, \\ \text{subject to} \ \ Ax = b, \ \ x \geqslant 0, \end{cases} \tag{1.7}$$

where $A \in \mathbb{R}^{m \times n}$, $b \in \mathbb{R}^m$ and $c \in \mathbb{R}^n$, and $x \geqslant 0$ stands for componentwise inequalities. It is clear that the above LP problem is a special case of NLP problem in which

$$f(x) = c^\top x, \quad g(x) = Ax - b, \quad h(x) = -x.$$

Another interesting problem is the so-called *quadratic programming* problem (QP problem, for short) which can be stated as follows:

$$\begin{cases} \min \ \dfrac{1}{2}x^\top Q x + c^\top x, \\ \text{subject to} \ \ Ax = a, \ \ Bx \leqslant b, \end{cases} \tag{1.8}$$

where Q, A, B are certain matrices, and a, b, c are certain vectors. This is a special case of the NLP problem with

$$f(x) = \frac{1}{2}x^\top Q x + c^\top x, \quad g(x) = Ax - a, \quad h(x) = Bx - b.$$

Let us now look at the case of single-variable functions. The following result is concerned with local extremes of single-variable functions.

Theorem 1.1. Let $f : (a, b) \to \mathbb{R}$.

(i) If $f(\cdot)$ attained a local extreme at $x_0 \in (a, b)$ and $f'(x_0)$ exists, then

$$f'(x_0) = 0. \tag{1.9}$$

(ii) If $f(\cdot)$ is C^1 near $x_0 \in (a, b)$, and (1.9) holds, then

- x_0 is a local minimum of $f(\cdot)$ if

$$f'(x_0 - \delta) < 0 < f'(x_0 + \delta), \quad \text{for all sufficiently small } \delta > 0;$$

- x_0 is a local maximum of $f(\cdot)$ if

$$f'(x_0 - \delta) > 0 > f'(x_0 + \delta), \quad \text{for all sufficiently small } \delta > 0;$$

- x_0 is not a local extreme of $f(\cdot)$ if there exists a $\delta > 0$ such that

$$f'(x_1)f'(x_2) > 0, \qquad \forall 0 < |x_1 - x_0|, |x_2 - x_0| < \delta.$$

(iii) If $f(\cdot)$ is C^2 near x_0, and $x_0 \in (a, b)$ is a local minimum (resp. a local maximum) of $f(\cdot)$, then (1.9) holds and

$$f''(x_0) \geqslant 0. \qquad (\text{resp. } f''(x_0) \leqslant 0)$$

(iv) If $x_0 \in (a, b)$ such that (1.9) holds and

$$f''(x_0) > 0. \qquad (\text{resp. } f''(x_0) < 0)$$

Then x_0 is a local minimum (resp. local maximum) of $f(\cdot)$.

In the above theorem, Part (i) is called the *Fermat's Theorem* (see Theorem 4.2 in Chapter 1), which is also called the *first order necessary condition* for local extremes; Part (ii) is called the *First Order Derivative Test*; Part (iii) is called the *second order necessary condition* for local extremes; and Part (iv) is called the *Second Order Derivative Test*. Note that when (1.9) holds and

$$f''(x_0) = 0, \tag{1.10}$$

no conclusion can be drawn. In other words, the point x_0 could be either a local maximum, a local minimum, or neither.

Next, we look at the following result which is concerned with global extreme of single variable functions.

Proposition 1.2. (i) *Let* $f : [a, b] \to \mathbb{R}$ *be continuous and piecewise* C^1, *i.e., there are finitely many points* $a = a_0 < a_1 < \cdots < a_m = b$ *such that* $f(\cdot)$ *is* C^1 *on each* (a_i, a_{i+1}) $(0 \leqslant i \leqslant m - 1)$. *Let*

$$\mathcal{K} = \{x \in (a, b) \mid f'(x) = 0\} \bigcup \{a_0, \cdots, a_m\}.$$

Then

$$\min f([a, b]) = \min f(\mathcal{K}), \qquad \max f([a, b]) = \max f(\mathcal{K}).$$

(ii) *Let* $f : \mathbb{R} \to \mathbb{R}$ *be continuous and piecewise* C^1 *on* $\mathbb{R}$, *i.e., there are finitely many points* $-\infty = a_0 < a_2 < \cdots < a_m = \infty$ *such that* $f(\cdot)$ *is* C^1 *on each* (a_i, a_{i+1}) $(0 \leqslant i \leqslant m - 1)$. *Let* $f(\cdot)$ *be coercive, i.e.,*

$$\lim_{\|x\| \to \infty} f(x) = \infty, \tag{1.11}$$

and

$$\mathcal{K} = \{x \in \mathbb{R} \mid f'(x) = 0\} \bigcup \{a_1, \cdots, a_{m-1}\}.$$

Then

$$\min f([a, b]) = \min f(\mathcal{K}).$$

The proof follows from the Weierstrass Theorem and the Fermat Theorem. Note that usually, $\mathcal{K}$ is a finite set and evaluating $f(\cdot)$ over $\mathcal{K}$ is not very difficult. Therefore, the above gives a construction of global minimum and maximum for the function $f(\cdot)$ over $[a, b]$. Note that the case $m = 1$ is well-known in standard calculus courses.

We now present a couple of examples.

Example 1.3. Consider

$$f(x) = x^2 - 4|x| - 2x, \qquad x \in \mathbb{R}.$$

This function is continuous on $\mathbb{R}$ and differentiable at each point of $\mathbb{R} \setminus \{0\}$. We have

$$f(x) = \begin{cases} x^2 + 2x, & x \in (-\infty, 0), \\ x^2 - 6x, & x \in [0, \infty). \end{cases}$$

Hence,

$$f'(x) = \begin{cases} 2x + 2, & x \in (-\infty, 0), \\ 2x - 6, & x \in (0, \infty). \end{cases}$$

Consequently, there are three critical points $x = -1, 0, 3$. On the other hand, $f(x)$ is coercive. Since

$$f(-1) = -1, \quad f(0) = 0, \quad f(3) = -9,$$

one has

$$\min_{x \in \mathbb{R}} f(x) = f(-3) = -9.$$

Example 1.4. Let

$$f(x) = 2(x + 1)^2 + |x - 1| - |2x + 1|, \qquad x \in \mathbb{R}.$$

Then $f(\cdot)$ is piecewise differentiable. We actually have

$$f(x) = \begin{cases} 2(x+1)^2 + x - 1 - (2x+1) = 2(x+1)^2 - x - 2, & x \geqslant 1, \\ 2(x+1)^2 + 1 - x - (2x+1) = 2(x+1)^2 - 3x, & -\dfrac{1}{2} \leqslant x \leqslant 1, \\ 2(x+1)^2 + 1 - x + 2x + 1 = 2(x+1)^2 + x + 2, & x \leqslant -\dfrac{1}{2}. \end{cases}$$

Consequently,

For $x \geqslant 1$, $\quad 0 = f'(x) = 4x + 3$, $\quad$ not possible.

For $-\frac{1}{2} \leqslant x \leqslant 1$, $\quad 0 = f'(x) = 4x + 1$, $\quad \Rightarrow \quad x = -\dfrac{1}{4}$,

For $x \leqslant -\frac{1}{2}$, $\quad f'(x) = 4x + 5$, $\quad \Rightarrow \quad x = -\dfrac{5}{4}$.

We evaluate the following:

$$f(1) = 5, \quad f\left(-\frac{1}{2}\right) = 2,$$

$$f\left(-\frac{1}{4}\right) = \frac{15}{8}, \quad f\left(-\frac{5}{4}\right) = \frac{7}{8}.$$

Hence, noting the coercivity of $f(x)$, we obtain

$$\min f(x) = f\left(-\frac{5}{4}\right) = \frac{7}{8}, \qquad \max f(x) = f(1) = 5.$$

Example 1.5. Let

$$f(x) = |\sin x| + |\cos x|, \qquad x \in [0, 2\pi].$$

Then

$$f(x) = \begin{cases} \sin x + \cos x, & x \in \left[0, \frac{\pi}{2}\right], \\ \sin x - \cos x, & x \in \left[\frac{\pi}{2}, \pi\right], \\ -\sin x - \cos x, & x \in \left[\pi, \frac{3\pi}{2}\right], \\ -\sin x + \cos x, & x \in \left[\frac{3\pi}{2}, 2\pi\right]. \end{cases}$$

Now, we have the following:

$$\text{On } \left[0, \frac{\pi}{2}\right], \qquad 0 = f'(x) = \cos x - \sin x \qquad \Rightarrow \qquad x = \frac{\pi}{4};$$

$$\text{On } \left[\frac{\pi}{2}, \pi\right], \qquad 0 = f'(x) = \cos x + \sin x \qquad \Rightarrow \qquad x = \frac{3\pi}{4};$$

$$\text{On } \left[\pi, \frac{3\pi}{2}\right], \qquad 0 = f'(x) = -\cos x + \sin x \qquad \Rightarrow \qquad x = \frac{5\pi}{4};$$

$$\text{On } \left[\frac{3\pi}{2}, 2\pi\right], \qquad 0 = f'(x) = -\cos x - \sin x \qquad \Rightarrow \qquad x = \frac{7\pi}{4}.$$

We now evaluate:

$$f(0) = f\left(\frac{\pi}{2}\right) = f(\pi) = f\left(\frac{3\pi}{2}\right) = f(2\pi) = 1,$$

$$f\left(\frac{\pi}{4}\right) = f\left(\frac{3\pi}{4}\right) = f\left(\frac{5\pi}{4}\right) = f\left(\frac{7\pi}{4}\right) = \sqrt{2}.$$

Hence, we obtain

$$\max f(x) = \sqrt{2}, \qquad \min f(x) = 1.$$

Exercises

1. Find examples showing that for some $G \subseteq \mathbb{R}$ and $f : G \to \mathbb{R}$, although $f(\cdot)$ attains a global extreme at $x_0 \in G$, but $f'(x_0) \neq 0$.

2. Give some examples $f : (a, b) \to \mathbb{R}$ with $x_0 \in (a, b)$ such that $f'(x_0) = f''(x_0) = 0$, but no definite conclusions can be drawn if x_0 is a local extreme of $f(\cdot)$.

3. For given function $f(x)$ and set G, find the global maximum and global minimum of $f(x)$ over G:

(i) $f(x) = x^3 - 2x^2 + x - 1$, and $G = [0, 2]$;

(ii) $f(x) = \ln(x^2 + 1) - \frac{1}{5}x^2$, $G = [-1, 3]$;

(iii) $f(x) = \sin^2 x + \cos x$, and $G = [0, 2\pi]$.

4. Let $f(x) = |x| + (x - 1)^2$. Find the minimum of $f(x)$.

5*. Let $f(x, \lambda, a) = \lambda|x| + (x - a)^2$ with $\lambda, a \in \mathbb{R}$ being parameters. For any $(\lambda, a) \in \mathbb{R}^2$, find the minimum of $f(x, \lambda, a)$.

6*. Let $f(x) = \sum_{i=1}^{n} \lambda_i|x - a_i| + \sum_{i=1}^{n} \mu_i(x - b_i)^2$, where $\lambda_i, \mu_i, a_i \in \mathbb{R}$, $i = 1, 2, \cdots, n$. Find the minimum of $f(x)$ over $\mathbb{R}$.

7*. Prove Proposition 1.2.

2 Some Examples of Optimization Problems

In this section, we present some examples of optimization problems.

Example 2.1. (Distance) Let

$$F_1 = \{(x, y) \in \mathbb{R}^2 \mid (x - 1)^2 + (y + 1)^2 = 1\},$$
$$F_2 = \{(x, y) \mid y \geq 1 + x^2\}.$$

Find the distance between F_1 and F_2, i.e., find $(\bar{x}_1, \bar{y}_1) \in F_1$ and $(\bar{x}_2, \bar{y}_2) \in F_2$ such that

$$(\bar{x}_1 - \bar{x}_2)^2 + (\bar{y}_1 - \bar{y}_2)^2 = \inf_{\substack{(x_1, y_1) \in F_1 \\ (x_2, y_2) \in F_2}} \left[(x_1 - x_2)^2 + (y_1 - y_2)^2\right].$$

If we let $G = \mathbb{R}^4$,

$$F = F_1 \times F_2 \equiv \{(x_1, y_1, x_2, y_2) \in \mathbb{R}^4 \mid (x_1, y_1) \in F_1, \ (x_2, y_2) \in F_2\}$$

and

$$f(x_1, y_1, x_2, y_2) = (x_1 - x_2)^2 + (y_1 - y_2)^2, \qquad (x_1, y_1, x_2, y_2) \in G,$$

then the above problem becomes Problem (G).

Example 2.2. (Approximation) Let $g : [a, b] \to \mathbb{R}$ be a continuous function whose values $g(x_k)$ at x_k ($0 \leqslant k \leqslant m$) are known. We want to find a polynomial of degree $n < m$ which approximates $g(\cdot)$ in a certain sense. To formulate the problem as an optimization problem, we let any general polynomial of degree n be written as follows:

$$p(x) = a_n x^n + a_{n-1} x^{n-1} + \cdots + a_1 x + a_0.$$

Define

$$\varepsilon_k(a_0, a_1, \cdots, a_n) = g(x_k) - p(x_k) = g(x_k) - \sum_{i=0}^{n} a_i x_k^i, \quad 0 \leqslant k \leqslant m.$$

We want to choose $a_0, a_1, \cdots, a_n$ such that the following is minimized:

$$
\begin{aligned}
f(a_0, a_1, \cdots, a_n) &= \sum_{k=0}^{m} \varepsilon_k(a_0, a_1, \cdots, a_n)^2 \\
&= \sum_{k=0}^{m} \left[g(x_k) - \sum_{i=0}^{n} a_i x_k^i \right]^2 \\
&= \sum_{k=0}^{m} \left[g(x_k)^2 - 2g(x_k) \sum_{i=0}^{n} a_i x_k^i + \left(\sum_{i=0}^{n} a_i x_k^i \right)^2 \right] \\
&= \sum_{i,j=0}^{n} \left(\sum_{k=0}^{m} x_k^{i+j} \right) a_i a_j - 2 \sum_{i=0}^{n} \left(\sum_{k=0}^{m} g(x_k) x_k^i \right) a_i + \sum_{k=0}^{m} g(x_k)^2 \\
&\equiv a^\top Q a - 2 q^\top a + c,
\end{aligned}
$$

for obvious definitions of Q, q, and c. Hence, we are looking for $a \in \mathbb{R}^{n+1}$ such that the above quadratic function is minimized. Further, if we want all $a_i \geqslant 0$, then we may further impose constraint

$$a \geqslant 0.$$

Example 2.3. (Utility maximization) Let a consumer consume n commodities with the amount of the i-th commodity being $x_i \geqslant 0$. The satisfaction level of consuming x_i units of the i-th commodity ($1 \leqslant i \leqslant n$) can be described by the value $u(x) \equiv u(x_1, \cdots, x_n)$ of a *utility function* $u : \mathbb{R}_+^n \to \mathbb{R}$. Suppose the price for the i-th commodity is $p_i > 0$, and the consumer has a total budget of $a > 0$. Therefore, we have the following constraint, with $p = (p_1, \cdots, p_n)^\top \in \mathbb{R}^n_{++} \equiv \{p \in \mathbb{R}^n \mid p > 0\}$,

$$p^\top x \leqslant a.$$

Now, we define

$$F = \left\{x \in \mathbb{R}^n_+ \mid p^\top x \leqslant a\right\},$$

and pose Problem (G′) with the above F and with objective function $f(x) = u(x)$. One concrete example of $u(x)$ takes the following form, which is called the *Cobb–Douglas* function:

$$u(x_1, \cdots, x_n) = (x_1)^{a_1} \cdots (x_n)^{a_n}, \qquad a_i > 0,\ 1 \leqslant i \leqslant n, \qquad \sum_{i=1}^n a_i = 1.$$

Although there are no strict restrictions, rational smooth utility functions $u(\cdot)$ satisfies

$$u_x(x) > 0, \qquad u_{xx}(x) < 0.$$

In the case $n = 1$, the first in the above means "the more the better" and the second in the above means "the marginal utility is decreasing."

Example 2.4. (Profit maximization) Let a firm produce a product which needs n types of material. Let $y = g(x) \equiv g(x_1, \cdots, x_n)$ be the number of the product which can be produced by using x_i units of the i-th material. We call $g : \mathbb{R}^n_+ \to \mathbb{R}_+$ a *production function*. Let w^i be the price of the i-th material, $w = (w_1, \cdots, w_n)^\top \in \mathbb{R}^n_+$, and let $p : \mathbb{R}_+ \to \mathbb{R}_+$ be the *demand function* where $p(y)$ is the price of the product when the supply level is y. Thus, $p(y)$ is non-increasing function: the more the supply, the lower the price. Thus, the total profit will be

$$f(x) = p(g(x))g(x) - w^\top x.$$

Then we can pose Problem (G′) with the above $f(\cdot)$ and with $F = \mathbb{R}^n_+$.

Example 2.5. (Cost minimization) Let $g(x)$ be a production function, and a firm needs to meet the minimum production level: $g(x) \geqslant \bar{y}$ for a fixed $\bar{y} > 0$. The price for the i-th material is w^i. The firm would like to minimize the total cost

$$f(x) = w^\top x,$$

subject to the constraint $g(x) \geqslant \bar{y}$. This is a special case of Problem (G).

Example 2.6. (Portfolio selection) Suppose in a security market there are $n+1$ assets, the 0-th asset is a bond which is riskless meaning that the value of the amount invested in will always be non-decreasing, and the last n assets are stocks which are risky meaning that the value of the amount invested might be increasing or decreasing. Let the current price of the i-th stock be $\bar{w}_i$ $(1 \leqslant i \leqslant n)$ and the bond price be w_0. Suppose there are m possible states of the world for a certain future time moment T, denoted by $\omega_1, \cdots, \omega_m$. If the state ω_j happens at time T, the i-th asset will have price $w_{ij} > 0$. An agent has his initial wealth $y_0 > 0$ dollars in cash. He spends all the cash to buy x_i shares of the i-th asset so that

$$y_0 = \sum_{i=0}^{n} w_i x_i = w^\top x.$$

Then at time T, if ω_j happens, his wealth becomes

$$y_j(T) = \sum_{i=0}^{n} w_{ij} x_i.$$

Now, let the probability of ω_j be $p_j > 0$ $\left(\sum_{j=1}^{m} p_j = 1\right)$. Then the *expected utility* at time T is given by

$$f(x) = f(x_1, \cdots, x_n) = \sum_{j=1}^{m} p_j u\left(\sum_{i=0}^{n} w_{ij} x_i\right),$$

where $u(\cdot)$ is a *utility function*, measuring the satisfaction level. Then letting

$$F = \left\{ x \in \mathbb{R}_+^{n+1} \mid w^\top x = y_0 \right\},$$

we can pose Problem (G') for the above $f(\cdot)$ and F.

3   Existence of Optimal Solutions

In this section, we are going to look at the existence of optimal solutions to Problem (G). We hope to find conditions as weak as possible, yet the existence of optimal solution is guaranteed.

Recall that $F \subseteq G \subseteq \mathbb{R}^n$ and $f : G \to \mathbb{R}$. Next, we introduce the following definition.

Definition 3.1. (i) Problem (G) is said to be *finite*, if

$$\bar{f} = \inf_{x \in F} f(x) > -\infty. \tag{3.1}$$

(ii) Problem (G) is said to be (*uniquely*) *solvable* if there exists an (a unique) $\bar{x} \in F$ such that

$$\inf f(F) = f(\bar{x}). \tag{3.2}$$

Note that if Problem (G) is not finite, i.e., (3.1) fails, then optimal solution will not exist and this is the end of the story. For example, if $F = (0,1)$ and $f(x) = -\frac{1}{x}$. Then the corresponding Problem (G) is not finite, and no minimum will exist for $f(\cdot)$ over F. Finiteness of Problem (G) is usually very easy to check. Here is a proposition which collects some useful and easy-checking conditions for finiteness.

Proposition 3.2. *Let $F \subset \mathbb{R}^n$ and $f : F \to \mathbb{R}$. Then Problem (G) will be finite if one of the following holds:*

(i) $f(\cdot)$ *is bounded from below, i.e., there exists a constant $K > 0$ such that*

$$f(x) \geqslant -K, \qquad \forall x \in F. \tag{3.3}$$

(ii) *F is bounded, $f(\cdot)$ is continuous, and*

$$\varliminf_{x \to y} f(x) > -\infty, \qquad \forall y \in \partial F. \tag{3.4}$$

Proof. (i) is trivial.

(ii) Denote

$$g(y) = \varliminf_{x \to y} f(x), \qquad \forall y \in \partial F.$$

Then $g : \partial F \to \mathbb{R}$ is a well-defined function. By the definition of liminf, for any $y \in \partial F$, there exists a $\delta(y) > 0$ such that

$$f(x) > g(y) - 1, \qquad \forall x \in B_{\delta(y)}(y).$$

Since F is bounded, ∂F is compact. Hence, there are finitely many $y_1, \cdots, y_k \in \partial F$ such that

$$\partial F \subseteq \bigcup_{i=1}^{k} B_{\delta(y_i)}(y_i).$$

Then

$$f(x) \geqslant \min_{1 \leqslant i \leqslant k} g(y_i) - 1, \qquad \forall x \in \bigcup_{i=1}^{k} B_{\delta(y_i)}(y_i) \cap F \equiv F_0.$$

On the other hand, $\overline{F \setminus F_0}$ is compact. Thus, $f(\cdot)$ is bounded on $\overline{F \setminus F_0}$. This shows that $f(\cdot)$ is bounded from below on F. ∎

We point out that finiteness does not imply the solvability in general. Here is a simple example.

Example 3.3. Let $f(x) = e^{-|x|}$, $x \in \mathbb{R}$. Then

$$\inf_{x \in \mathbb{R}} f(x) = 0,$$

which implies that the corresponding Problem (G) is finite. But, the infimum is not attained.

On the other hand, we will see various cases that under certain conditions, finiteness could imply the solvability. The following is a basic general existence theorem for optimal solutions to Problems (G) and (G′).

Theorem 3.4. (Weierstrass) *Let F be compact and $f : F \to \mathbb{R}$ be continuous. Then both Problems (G) and (G′) admit optimal solutions.*

This is nothing but Part (i) of Theorem 3.3 in Chapter 1.

Example 3.5. Let

$$f(x, y, z) = x^2 - y^3 + \sin z, \qquad (x, y, z) \in \mathbb{R}^3.$$

$$F = \left\{ (x, y, z) \mid x^4 + y^4 + z^4 \leqslant 1 \right\}.$$

Then F is compact. Hence, f admits a minimum and maximum over F, by Weierstrass' Theorem.

The following is an extension of above Weierstrass' Theorem, in which conditions are relaxed.

Theorem 3.6. *Let $F \subseteq \mathbb{R}^n$ and $f : F \to \mathbb{R}$ be lower semi-continuous such that*

$$\inf_{x \in F} f(x) > -\infty. \tag{3.5}$$

Further, there exists a constant $R > 0$ such that $F \cap \bar{B}_R(0)$ is compact and

$$\inf_{\|x\| > R, \ x \in F} f(x) > \inf_{x \in F} f(x). \qquad (\inf_{\varnothing} f(x) \triangleq \infty). \tag{3.6}$$

Then Problem (G) admits an optimal solution.

Proof. From (3.6), we see that

$$\inf_{x\in F} f(x) = \inf_{x\in F\cap \bar{B}_R(0)} f(x).$$

Therefore, we can find a minimizing sequence $x_k \in F \cap \bar{B}_R(0)$. By the compactness of $F \cap \bar{B}_R(0)$, we may assume that $x_k \to \bar{x} \in F \cap \bar{B}_R(0)$. Then by the lower semi-continuity of $f(\cdot)$, one has

$$f(\bar{x}) \leqslant \lim_{k\to\infty} f(x_k) = \inf_{x\in F} f(x).$$

Hence, $\bar{x} \in F$ has to be an optimal solution to Problem (G). $\qquad\square$

Corollary 3.7. *Suppose $F = \mathbb{R}^n$ and $f : \mathbb{R}^n \to \mathbb{R}$ is lower semi-continuous. Suppose*

$$\lim_{\|x\|\to\infty} f(x) > \inf_{x\in\mathbb{R}^n} f(x) > -\infty, \tag{3.7}$$

which is the case if

$$\lim_{\|x\|\to\infty} f(x) = \infty. \tag{3.8}$$

Then Problem (G) admits an optimal solution.

Proof. Condition (3.7) implies condition (3.5)–(3.6). $\qquad\square$

When $f(\cdot)$ satisfies (3.8), we say that $f(\cdot)$ is *coercive*. For Problem (G'), we can have a similar result. The details are left to the readers.

The following example shows that sometimes, Corollary 3.7 could be very useful.

Example 3.8. Let $G = F = \mathbb{R}^2$,

$$f(x,y) = \sup_{k\geqslant 1}\Big[\sin^k(x+y) + x^2 + y^2 \Big], \qquad (x,y) \in \mathbb{R}^2.$$

The problem is to minimize $f(x,y)$ subject to $(x,y) \in \mathbb{R}^2$. Let

$$g(z) = \sup_{k\geqslant 1}\Big[\sin^k z \Big], \qquad z \in \mathbb{R}.$$

Then $g(\cdot)$ is lower semi-continuous (see Proposition 3.7 of Chapter 1). Hence,

$$f(x,y) = g(x+y) + x^2 + y^2$$

is lower semi-continuous on $\mathbb{R}^2$. Further, it is coercive. Therefore, the corresponding Problem (G) admits a minimum.

For (1.4), we have the following result.

Proposition 3.9. *Let $f : \mathbb{R}^n \to \mathbb{R}$, $g : \mathbb{R}^n \to \mathbb{R}^m$ and $h : \mathbb{R}^n \to \mathbb{R}^\ell$ be all continuous such that the feasible set F defined by (1.5) is non-empty. Suppose either F is bounded or $f(\cdot)$ is coercive. Then Problem (1.4) admits an optimal solution.*

Note that in the above proposition, the set F defined by (1.5) is always closed. Hence, boundedness of F implies its compactness. Practically, for Problem (1.4), it is easy to check either the boundedness of F or the coercivity of $f(\cdot)$. Therefore, Proposition 3.9 is easy to use to get the existence of optimal solutions.

Example 3.10. Consider following quadratic function

$$f(x) = x^\top Q x + 2c^\top x, \qquad x \in \mathbb{R}^n, \tag{3.9}$$

with $Q \in \mathbb{S}^n$ being positive definite. Then there exists a $\delta > 0$ such that

$$f(x) = x^\top Q x + 2c^\top x \geqslant \delta \|x\|^2 - 2\|c\| \, \|x\| = \delta \left(\|x\| - \frac{\|c\|}{2\delta} \right)^2 - \frac{\|c\|^2}{4\delta}.$$

Hence,

$$\lim_{\|x\| \to \infty} f(x) = \infty.$$

By Corollary 3.7, there exists a minimum of $f(x)$ over any closed set in $\mathbb{R}^n$, including $\mathbb{R}^n$ itself.

Actually, for quadratic function, we have the following interesting result.

Proposition 3.11. *Let $Q \in \mathbb{S}^n$ and $c \in \mathbb{R}^n$ and let $f(\cdot)$ be defined by (3.9). Then the following are equivalent:*

(i) *Function $f(\cdot)$ admits a minimum over $\mathbb{R}^n$.*

(ii) *Function $f(\cdot)$ is bounded from below.*

(iii) *$c \in \mathcal{R}(Q)$, and $Q \geqslant 0$.*

The proof is left to the readers. The point of the above proposition is that for quadratic function $f(\cdot)$ defined by (3.9), bounded from below implies the existence of a minimum, which is essentially different from Example 3.3.

Example 3.12. Let

$$f(x, y) = x^2 + y^3, \qquad (x, y) \in \mathbb{R}^2,$$

and

$$F = \{(x, y) \in \mathbb{R}^2 \mid y \geqslant -1\}.$$

Note that although $f(x, y)$ is not coercive over $\mathbb{R}^2$, it is coercive over F in the sense that

$$\lim_{(x,y) \in F, \, \|(x,y)\| \to \infty} f(x, y) = +\infty.$$

Then Proposition 3.6 applies to get the existence of optimal solutions for the corresponding Problem (G).

We now look at an important and interesting special case that $f(x)$ is linear, i.e.,

$$f(x) = a^\top x + b, \qquad x \in \mathbb{R}^n, \tag{3.10}$$

for some $a \in \mathbb{R}^n \setminus \{0\}$ and $b \in \mathbb{R}$. We have the following result.

Proposition 3.13. *Let $F \subseteq \mathbb{R}^n$ be non-empty and closed, and $f(x)$ be given by (3.10).*

(i) *If*

$$\inf_{x \in F} f(x) > -\infty,$$

then for any $x \in F$ and any $c \in \mathbb{R}^n$ with $a^\top c > 0$, there exists a $\lambda(x, c) > 0$ such that

$$x - \lambda c \notin F, \qquad \forall \lambda \geqslant \lambda(x, c). \tag{3.11}$$

(ii) *If for any $x \in F$, there exists a $c \in \mathbb{R}^n$ with $a^\top c \geqslant 0$ such that for some $\lambda \geqslant 0$,*

$$x - \lambda c \in \partial F,$$

then

$$\inf_{x \in F} f(x) = \inf_{x \in \partial F} f(x). \tag{3.12}$$

In particular, if F is compact, then

$$\min_{x \in F} f(x) = \min_{x \in \partial F} f(x). \tag{3.13}$$

(iii) *If there exists a constant $K > 0$ such that*

$$F_K \equiv \{x \in F \mid a^\top x \leqslant K\}$$

is non-empty and bounded, then $f(\cdot)$ has a minimum on F such that (3.13) holds.

Proof. (i) If the conclusion is false, then there exist an $\bar{x} \in F$ and a $c \in \mathbb{R}^n$ with $a^\top c > 0$ such that

$$\bar{x} - \lambda c \in F, \qquad \forall \lambda > 0.$$

This will lead to

$$f(\bar{x} - \lambda c) = a^\top (\bar{x} - \lambda c) + b = f(\bar{x}) - \lambda a^\top c \to -\infty, \qquad \lambda \to \infty,$$

a contradiction.

(ii) For any $x \in F$, we have some $c \in \mathbb{R}^n$ with $a^\top c \geqslant 0$ and some $\lambda \geqslant 0$ such that $x - \lambda c \in \partial F$. Thus,

$$\inf_{y \in \partial F} f(y) \leqslant f(x - \lambda c) = f(x) - \lambda a^\top c \leqslant f(x), \qquad \forall x \in F,$$

which leads to

$$\inf_{y \in \partial F} f(y) \leqslant \inf_{x \in F} f(x) \leqslant \inf_{x \in \partial F} f(x),$$

proving (3.12). Finally, if F is compact, then ∂F is also compact and (3.12) becomes (3.13).

(iii) By the definition of F_K, we see that by picking any $\bar{x} \in F_K$, one has

$$f(\bar{x}) = a^\top \bar{x} + b \leqslant K + b < a^\top x + b = f(x), \qquad \forall x \in F \setminus F_K.$$

Thus,

$$\inf_{x \in F} f(x) = \inf_{x \in F_K} f(x).$$

Further, by assumption, F_K is closed and bounded. Hence, Weierstrass' Theorem applies. □

The above result can be used to solve some cases of Problem (G). Here is an example.

Example 3.14. Let

$$F = \big\{ (x, y) \in \mathbb{R}^2 \mid x, y \geqslant 0, \ x + y \leqslant 1 \big\},$$

and

$$f(x, y) = x - y.$$

Clearly, F is a non-empty compact set. Hence, by Weierstrass' Theorem, $f(x, y)$ admits a minimum over F. By Proposition 3.13 (iii), there must also be a minimum $(\bar{x}, \bar{y})$ appearing on the boundary ∂F. Thus, we have three cases:

(i) $\bar{x} = 0$, $0 \leqslant \bar{y} \leqslant 1$. If this case happens, one must have

$$f(\bar{x}, \bar{y}) = -\bar{y} \geqslant -1 = f(0, 1).$$

Hence, we must have $(\bar{x}, \bar{y}) = (0, 1)$.

(ii) $\bar{y} = 0$, $0 \leqslant \bar{x} \leqslant 1$. If this case happens, one must have

$$f(\bar{x}, \bar{y}) = \bar{x} \geqslant 0 > -1 = f(0, 1).$$

This means that case (ii) will not happen.

(iii) $\bar{x} + \bar{y} = 1$, $\bar{x}, \bar{y} \geqslant 0$. If this case happens, we have

$$f(\bar{x}, \bar{y}) = \bar{x} - (1 - \bar{x}) = 2\bar{x} - 1 \geqslant -1 = f(0, 1).$$

Combining the above cases, we see that the minimum of $f(x, y)$ must be attained at $(0, 1)$, with the minimum value of $f(0, 1) = -1$.

In the case that F might not be compact, we present the following example.

Example 3.15. Let

$$F = \{(x, y) \in \mathbb{R}^2 \mid x, y \geqslant 0, \ x + y \geqslant 1\},$$

and

$$f(x, y) = 2x + 3y, \qquad (x, y) \in \mathbb{R}^2.$$

Clear, $f(x, y)$ is a linear function. However, F is not compact. But, for any $(x, y) \in F$, one has

$$f(x, y) = 2x + 3y = 2(x + y) + y \geqslant 2 + y \geqslant 2 = f(1, 0),$$

where $(1, 0) \in F$. Hence, $f(x, y)$ has a minimum over F. Note that for this example, condition in (iii) of Proposition 3.13 holds.

It is obvious that for maximization problem of linear functions, we have similar results as Propositions 3.13. The details are left to the interested readers.

Exercises

1. Let $f(x, y, z) = x^2 + 2y^2 + 3z^2 + x - 3y - 5z$. Show that $f(x, y, z)$ admits a minimum over $\mathbb{R}^3$. Does it have a maximum over $\mathbb{R}^3$? Why?

2. Let $f(x, y) = \frac{1}{x^2} + \frac{1}{y^2} + x^4 + y^4$ and $F = \{(x, y) \in \mathbb{R}^2 \mid x, y \neq 0\}$. Show that $f(x, y)$ has a minimum over F.

3. Let $F = \{(x,y) \in \mathbb{R}^2 \mid y \geqslant x^2\}$ and $f(x,y) = -x^3 + y^2$. Determine if $f(x,y)$ has a minimum over F. Give your reason. How about maximum?

4. Let $F = \{(x,y) \in \mathbb{R}^2 \mid x, y \geqslant 0, \ x + y \leqslant 1\}$, and $f(x,y) = x^2 - y^2$. Show that $f(x,y)$ has a maximum and a minimum over F.

5*. State and prove a similar result as Proposition 3.6 for Problem (G′).

6*. Prove Proposition 3.11.

7*. Let $F_1 \subseteq \mathbb{R}^{n_1}$ and $F_2 \subseteq \mathbb{R}^{n_2}$ be non-empty with F_1 being compact (and no conditions are assumed for F_2). Let $F = F_1 \times F_2$ and $f : F \to \mathbb{R}$ be continuous and bounded. Further, let

$$g(x) = \sup_{y \in F_2} f(x,y), \qquad x \in F_1.$$

Show that $g(x)$ admits a minimum over F_1.

8. Let $F = \{(x,y) \in \mathbb{R}^2 \mid x \geqslant (y-1)^2\}$ and $f(x,y) = x + y$. Find a minimum of $f(x,y)$ over F.

9. Let $F = \{(x,y) \in \mathbb{R}^2 \mid x \geqslant 1, \ 0 \leqslant y \leqslant x\}$ and $f(x,y) = 4x + y + 1$. Find a minimum of $f(x,y)$ over F.

10. State and prove the results of Proposition 3.13 for maximizing linear functions.

4   A Constructive Method

In this section, we present a constructive method for finding an optimal solution to Problem (G). The result is stated as follows.

Theorem 4.1. *Suppose there exists a* $\lambda \in \mathbb{R}^n \setminus \{0\}$ *such that for some* $\bar{x} \in F$, *the following hold:*

$$\min_{x \in \mathbb{R}^n} \left(f(x) + \lambda^\top x \right) = f(\bar{x}) + \lambda^\top \bar{x}, \tag{4.1}$$

and

$$\max_{x \in F} \lambda^\top x = \lambda^\top \bar{x}. \tag{4.2}$$

Then $\bar{x}$ *is an optimal solution to the corresponding Problem (G).*

Proof. For any $x \in F$, from (4.1),

$$f(x) + \lambda^\top x \geqslant f(\bar{x}) + \lambda^\top \bar{x}.$$

This implies, making use of (4.2),

$$f(x) - f(\bar{x}) \geqslant \lambda^\top \bar{x} - \lambda^\top x \geqslant 0.$$

Hence, $\bar{x} \in F$ is an optimal solution to Problem (G). $\qquad\square$

The above is called the *zeroth order sufficient condition* for Problem (G), because no derivative is needed here. We now use Theorem 4.1 to solve a Problem (G).

Example 4.2. Consider Problem (G) with

$$F = \left\{ (x,y) \in \mathbb{R}^2 \mid (x-1)^2 + (y+2)^2 \leqslant 1 \right\}, \qquad f(x,y) = x^2 + y^2.$$

We need to find $\lambda = (\lambda_1, \lambda_2)$ so that one can find a point $(\bar{x}, \bar{y}) \in \mathbb{R}^2$ solving (4.1)–(4.2). To this end, we begin with (4.1). If $(\bar{x}, \bar{y})$ minimizes the following function over $\mathbb{R}^2$:

$$f(x,y) + \lambda^\top \begin{pmatrix} x \\ y \end{pmatrix} = x^2 + y^2 + \lambda_1 x + \lambda_2 y,$$

then one must have (by Fermat's Theorem)

$$\bar{x} = -\frac{\lambda_1}{2}, \qquad \bar{y} = -\frac{\lambda_2}{2}.$$

Next, from (4.2), we see that $(\bar{x}, \bar{y})$ should be a maximum of a linear function $\lambda_1 x + \lambda_2 y$ over F. Thus, we must have $(\bar{x}, \bar{y}) \in \partial F$. In fact, if $(\bar{x}, \bar{y}) \in F^\circ$, then by Fermat's Theorem, $(\lambda_1, \lambda_2) = 0$, a contradiction. Hence,

$$\left(-\frac{\lambda_1}{2} - 1 \right)^2 + \left(-\frac{\lambda_2}{2} + 2 \right)^2 = 1.$$

Note that F is in the fourth quadrant. Thus, $\lambda_1 \leqslant 0$ and $\lambda_2 \geqslant 0$. We claim that $\lambda_1 < 0$. Otherwise, $\bar{\lambda}_1 = 0$ implies $\bar{x} = 0$ and $\lambda_2 = 4$, $\bar{y} = -2$. Then (with $(x,y) = (1,-1) \in \partial F$)

$$\lambda_1 \bar{x} + \lambda_2 \bar{y} = 4\bar{y} = -8 < -4 = \lambda_1 \cdot 1 + \lambda_2 \cdot (-1),$$

which means that $(\bar{x}, \bar{y})$ is not a maximum of the linear function $\lambda_1 x + \lambda_2 y$ over the set F, a contradiction. By the way, $\lambda_2 > 0$ since F does not touch x-axis.

On the other hand, ∂F can be decomposed into two parts, respectively represented by the following:

$$(\partial F)_+ = \left\{ (x,y) \in \mathbb{R}^2 \mid y = -2 + \sqrt{1 - (x-1)^2}, \quad x \in [0,2] \right\},$$
$$(\partial F)_- = \left\{ (x,y) \in \mathbb{R}^2 \mid y = -2 - \sqrt{1 - (x-1)^2}, \quad x \in [0,2] \right\}.$$

Thus, the linear function $\lambda_1 x + \lambda_2 y$ restricted on $(\partial F)_\pm$ can be written as

$$g_\pm(x) = \lambda_1 x + \lambda_2\left(-2 \pm \sqrt{1 - (x-1)^2}\right), \qquad x \in [0,2],$$

To maximize the above two functions, we first look at $g_+(\cdot)$ (on $(\partial F)_+$). Set

$$0 = g'_+(\bar{x}) = \lambda_1 - \frac{\lambda_2(\bar{x}-1)}{\sqrt{1-(\bar{x}-1)^2}}.$$

Since $\lambda_1 < 0$ and $\lambda_2 > 0$, we must have $\bar{x} < 1$. The above also yields

$$\lambda_1^2\left(1 - (\bar{x}-1)^2\right) = \lambda_2^2(\bar{x}-1)^2,$$

which leads to

$$\bar{x} - 1 = \frac{\lambda_1}{\sqrt{\lambda_1^2 + \lambda_2^2}} < 0.$$

Then (noting that $(\bar{x}, \bar{y}) \in (\partial F)_+$)

$$\bar{y} + 2 = \sqrt{1 - (\bar{x}-1)^2} = \sqrt{1 - \frac{\lambda_1^2}{\lambda_1^2 + \lambda_2^2}} = \frac{\lambda_2}{\sqrt{\lambda_1^2 + \lambda_2^2}}.$$

Consequently,

$$-\frac{\lambda_1}{2} = \bar{x} = 1 + \frac{\lambda_1}{\sqrt{\lambda_1^2 + \lambda_2^2}}, \qquad -\frac{\lambda_2}{2} = \bar{y} = -2 + \frac{\lambda_2}{\sqrt{\lambda_1^2 + \lambda_2^2}}.$$

Thus,

$$-\frac{1}{2} - \frac{1}{\lambda_1} = \frac{1}{\sqrt{\lambda_1^+ \lambda_2^2}} = -\frac{1}{2} + \frac{2}{\lambda_2},$$

which leads to

$$\lambda_2 = -2\lambda_1.$$

Then

$$\bar{x} = 1 + \frac{\lambda_1}{\sqrt{\lambda_1^2 + \lambda_2^2}} = 1 + \frac{\lambda_1}{\sqrt{5\lambda_1^2}} = 1 - \frac{1}{\sqrt{5}} = \frac{\sqrt{5}-1}{\sqrt{5}}, \qquad (4.3)$$

and

$$\bar{y} = -2 + \frac{\lambda_2}{\sqrt{\lambda_1^2 + \lambda_2^2}} = -2 - \frac{2\lambda_1}{\sqrt{5\lambda_1^2}} = -2 + \frac{2}{\sqrt{5}} = -\frac{2(\sqrt{5}-1)}{\sqrt{5}}. \qquad (4.4)$$

Therefore,

$$\lambda_1 = -2\bar{x} = -\frac{2(\sqrt{5}-1)}{\sqrt{5}}, \qquad \lambda_2 = -2\bar{y} = \frac{4(\sqrt{5}-1)}{\sqrt{5}}, \qquad (4.5)$$

and

$$\max_{(x,y)\in(\partial F)_+} \left(\lambda_1 x + \lambda_2 y\right) = \lambda_1 \bar{x} + \lambda_2 \bar{y} = -2(\bar{x}^2 + \bar{y}^2)$$

$$= -2\left(\frac{(\sqrt{5}-1)^2}{5} + \frac{4(\sqrt{5}-1)^2}{5}\right) = -2(\sqrt{5}-1)^2.$$

Next, we look at $g_-(\cdot)$ (on $(\partial F)_-$). Set

$$0 = g'_-(\widehat{x}) = \lambda_1 + \frac{\lambda_2(\widehat{x}-1)}{\sqrt{1-(\widehat{x}-1)^2}}.$$

Again, since $\lambda_1 < 0$ and $\lambda_2 > 0$, we must have $\widehat{x} > 1$. The above also yields

$$\lambda_1^2\left(1-(\widehat{x}-1)^2\right) = \lambda_2^2(\widehat{x}-1)^2,$$

which leads to (note $\lambda_1 < 0$ and $\lambda_2 = -2\lambda_1$)

$$\widehat{x} - 1 = -\frac{\lambda_1}{\sqrt{\lambda_1^2 + \lambda_2^2}} = \frac{1}{\sqrt{5}}.$$

Then (noting that $(\widehat{x}, \widehat{y}) \in (\partial F)_-$)

$$\widehat{y} + 2 = -\sqrt{1-(\widehat{x}-1)^2} = -\sqrt{1-\frac{1}{5}} = -\frac{2}{\sqrt{5}}.$$

Hence,

$$\widehat{x} = \frac{\sqrt{5}+1}{\sqrt{5}}, \qquad \widehat{y} = -\frac{2(\sqrt{5}+1)}{\sqrt{5}},$$

and

$$\max_{(x,y)\in(\partial F)_-} \left(\lambda_1 x + \lambda_2 y\right) = \lambda_1 \widehat{x} + \lambda_2 \widehat{y} = -2(\widehat{x}^2 + \widehat{y}^2)$$

$$= -2\left(\frac{(\sqrt{5}+1)^2}{5} + \frac{4(\sqrt{5}+1)^2}{5}\right) = -2(\sqrt{5}+1)^2$$

$$< -2(\sqrt{5}-1)^2 = \max_{(x,y)\in(\partial F)_+} \left(\lambda_1 x + \lambda_2 y\right).$$

This means that the maximum of the linear function $(x, y) \mapsto \lambda_1 x + \lambda_2 y$ will be attained on $(\partial F)_+$. Now with (λ_1, λ_2) given by (4.5), the point $(\bar{x}, \bar{y})$ solves corresponding (4.1)–(4.2). Hence, $(\bar{x}, \bar{y})$ defined by (4.3)–(4.4) gives an optimal solution to the corresponding Problem (G).

In the above example, the function $(x, y) \mapsto x^2 + y^2$ is coercive on $\mathbb{R}^2$. The following example will show us how to treat a different situation.

Example 4.3. Consider Problem (G) with

$$F = \left\{(x, y) \in \mathbb{R}^2 \mid 0 \leqslant y \leqslant 1 - (x-1)^2, \ x \in [0, 2]\right\},$$

and

$$f(x,y) = x^2 - 2y^2, \qquad (x,y) \in \mathbb{R}^2.$$

To solve the problem, we first use Weierstrass' Theorem to get the existence of an optimal solution since F is compact and $f(x,y)$ is continuous. Note that unlike the previous example, $(x,y) \mapsto x^2 - 2y^2$ is not coercive over $\mathbb{R}^2$. Thus, the function $(x,y) \mapsto x^2 - 2y^2 + \lambda_1 x + \lambda_2 y$ (for some $\lambda_1, \lambda_2 \in \mathbb{R}$) will not have a global minimum over $\mathbb{R}^2$. To overcome this difficulty, we introduce the following function:

$$g(x,y) = x^2 + 2\varphi(y)y^2, \qquad (x,y) \in \mathbb{R}^2,$$

with $\varphi(\cdot)$ being a smooth function such that

$$\varphi(y) = \begin{cases} -1, & y \in [0,1], \\ 1, & y \in (-\infty, -1] \cup [2, \infty). \end{cases}$$

Then $(x,y) \mapsto g(x,y)$ is coercive in $\mathbb{R}^2$. Moreover, Problem (G) with $f(x,y)$ replaced by $g(x,y)$ is equivalent to the original Problem (G). We now use the method in the previous example to solve the problem.

For any $(\lambda_1, \lambda_2) \in \mathbb{R}^2$ given, since $(x,y) \mapsto g(x,y) + \lambda_1 x + \lambda_2 y$ is coercive on $\mathbb{R}^2$, it has a global minimum $(\bar{x}, \bar{y}) \in \mathbb{R}^2$. We require that $(\bar{x}, \bar{y}) \in F$. By Fermat's Theorem, we have

$$0 = 2\bar{x} + \lambda_1, \qquad 0 = 4\varphi(\bar{y})\bar{y} + 2\varphi'(\bar{y})\bar{y}^2 + \lambda_2 = -4\bar{y} + \lambda_2.$$

Then

$$0 \leqslant \bar{x} = -\frac{\lambda_1}{2} \leqslant 2, \qquad 0 \leqslant \bar{y} = \frac{\lambda_2}{4} \leqslant 1 - \left(-\frac{\lambda_1}{2} - 1\right)^2. \qquad (4.6)$$

We claim that $\lambda_1 < 0$. Otherwise, $\lambda_1 = 0$, which leads to $\lambda_2 = 0$. Hence, $\bar{x} = \bar{y} = 0$. However, if $0 < x < 1$, we have

$$x^2 < x \qquad \Longleftrightarrow \qquad x < 2x - x^2 = 1 - (x-1)^2.$$

This means $(x,x) \in F$, and

$$g(x,x) = f(x,x) = -x^2 < 0 = f(0,0) = g(0,0).$$

Thus, $(0,0)$ is not a minimum (of $g(x,y)$ over F), a contradiction. This implies

$$-4 \leqslant \lambda_1 < 0, \qquad \lambda_2 \geqslant 0.$$

Next, we need $(\bar{x}, \bar{y})$ to be a maximum of linear function $(x,y) \mapsto \lambda_1 x + \lambda_2 y$ over F. From the results of maximization of linear functions in the previous section, one must have $(\bar{x}, \bar{y}) \in \partial F$. Thus, it is the maximum of $\lambda_1 x + \lambda_2 y$ over ∂F. We write

$$\partial F = \Gamma_1 \cup \Gamma_2,$$

with

$$\Gamma_1 = \left\{ (x, 0) \mid x \in [0, 2] \right\}, \quad \Gamma_2 = \left\{ (x, y) \mid y = 1 - (x - 1)^2, \ x \in (0, 2) \right\}.$$

Now, let us look at two cases:

(i) If $(\bar{x}, \bar{y}) \in \Gamma_1$, we have $\bar{x} \in [0, 2]$ and $\bar{y} = 0$. Then for any $(x, y) \in \Gamma_1$, we should have

$$\lambda_1 x + \lambda_2 y = \lambda_1 x \leqslant \lambda_1 \bar{x} = -\frac{\lambda_1^2}{2}, \qquad \forall x \in [0, 2],$$

which leads to (taking $x = 0$ in the above)

$$0 \leqslant -\frac{\lambda_1^2}{2} < 0,$$

a contradiction. This means $(\bar{x}, \bar{y}) \notin \Gamma_1$.

(ii) Let $(\bar{x}, \bar{y}) \in \Gamma_2$. Then $(\bar{x}, \bar{y})$ is also a maximum of $\lambda_1 x + \lambda_2 y$ over Γ_2. For any $(x, y) \in \Gamma_2$, one has

$$\lambda_1 x + \lambda_2 y = \lambda_1 x + \lambda_2 [1 - (x - 1)^2], \qquad x \in (0, 2).$$

To maximize the above, we set (noting $\bar{x} = -\frac{\lambda_1}{2}$)

$$0 = \lambda_1 - 2\lambda_2(\bar{x} - 1) = \lambda_1 - 2\lambda_2\left(-\frac{\lambda_1}{2} - 1\right) = \lambda_1 + (\lambda_1 + 2)\lambda_2.$$

Thus, it is necessary that $\lambda_1 + 2 \neq 0$ (otherwise, the above leads to $\lambda_1 = 0$, a contradiction.). Hence,

$$\lambda_2 = -\frac{\lambda_1}{\lambda_1 + 2} > 0,$$

which further implies $\lambda_1 + 2 > 0$. Also, since $(\bar{x}, \bar{y}) \in \Gamma_2$, one obtains

$$\frac{\lambda_2}{4} = 1 - \left(-\frac{\lambda_1}{2} - 1\right)^2 = -\lambda_1 - \frac{\lambda_1^2}{4} = -\frac{\lambda_1(\lambda_1 + 4)}{4}$$

Then

$$-\frac{\lambda_1}{\lambda_1 + 2} = \lambda_2 = -\lambda_1(\lambda_1 + 4),$$

which leads to

$$0 = (\lambda_1 + 4)(\lambda_1 + 2) - 1 = \lambda_1^2 + 6\lambda_1 + 7.$$

Hence, noting $\lambda_1 + 2 > 0$,

$$\lambda_1 = \frac{-6 + \sqrt{36 - 28}}{2} = \frac{-6 + 2\sqrt{2}}{2} = -3 + \sqrt{2}.$$

Then

$$\lambda_2 = -\frac{\lambda_1}{\lambda_1 + 2} = -\frac{-3 + \sqrt{2}}{-1 + \sqrt{2}} = 1 + 2\sqrt{2}.$$

Consequently,

$$\bar{x} = -\frac{\lambda_1}{2} = \frac{-3 + \sqrt{2}}{2}, \qquad \bar{y} = \frac{\lambda_2}{4} = \frac{1 + 2\sqrt{2}}{4},$$

which gives an optimal solution to the original Problem (G).

From the above two examples, we see that Theorem 4.1 is applicable to Problem (G), with somehow complicated calculations. We will see that some more effective methods later.

Exercises

1. Let $F = \{(x, y) \in \mathbb{R}^2 \mid 2x^2 + y^2 \leqslant 1\}$ and $f(x, y) = x + y^2$. Find a minimum of $f(x, y)$ over F by using Theorem 4.1.

2. Let $F = \{(x, y) \in \mathbb{R}^2 \mid 1 \leqslant x^2 + y^2 \leqslant 4\}$ and $f(x, y) = x^2 + x - y^2$. Find a minimum of $f(x, y)$ over F by using Theorem 4.1.

3. Let $G \subseteq \mathbb{R}^n$ be a domain and $F \subseteq G$ be a compact set. Let $f : G \to \mathbb{R}$ and $g : G \to \mathbb{R}^m$ be continuous. Suppose there exists a $\lambda \in \mathbb{R}^m$ such that for some $\bar{x} \in F$, the following holds:

$$\min_{x \in G} \left[f(x) + \lambda^\top g(x) \right] = f(\bar{x}) + \lambda^\top g(\bar{x}),$$
$$\max_{x \in F} \lambda^\top g(x) = \lambda^\top g(\bar{x}).$$

Then $\bar{x}$ is an optimal solution of Problem (G).

4*. State and prove a result similar to Theorem 4.1 for Problem (G').

Chapter 3

Necessary and Sufficient Conditions of Optimal Solutions

Recall that the existence results of optimal solutions via Weierstrass type theorems presented in the previous chapter do not give constructions of optimal solutions. The constructive method introduced in Section 4 of the previous chapter is a little too complicated to use. Hence, we would like to establish some methods to solve the optimization problem more effectively.

1   Unconstrained Problems

When $F = \mathbb{R}^n$, we refer to Problem (G) as an unconstrained problem. We have the following result.

Proposition 1.1. *Let $F = \mathbb{R}^n$ and $\bar{x} \in \mathbb{R}^n$ be a local optimal solution to Problem (G).*

(i) *Suppose $f : F \to \mathbb{R}$ is Gâteaux differentiable at $\bar{x}$. Then*

$$f'(\bar{x}; u) \geqslant 0, \qquad \forall u \in \mathbb{R}^n. \tag{1.1}$$

(ii) *Suppose $f : F \to \mathbb{R}$ admits a gradient at $\bar{x}$. Then*

$$\nabla f(\bar{x}) = 0. \tag{1.2}$$

Proof. (i) Since $\bar{x} \in F$ is a local minimum point of $f(\cdot)$, for any $u \in \mathbb{R}^m$, we have

$$0 \leqslant \lim_{\delta \downarrow 0} \frac{f(\bar{x} + \delta u) - f(\bar{x})}{\delta} = f'(\bar{x}; u),$$

proving (i).

(ii) Suppose $f(\cdot)$ admits a gradient at $\bar{x}$. Then the above implies that

$$\nabla f(\bar{x})u \geqslant 0, \qquad \forall u \in \mathbb{R}^n,$$

which leads to (1.2). $\square$

Usually, the above is referred to as the *first order necessary condition*, which is a generalization of Fermat's theorem. Also, we see that in order part (ii) to be true, we need only the existence of a gradient and $f(\cdot)$ does not have to be (Fréchet) differentiable.

Example 1.2. Consider

$$f(x) = |x - 1| + x^2, \qquad x \in \mathbb{R}.$$

For any given $u \in \mathbb{R}$, when $x > 1$,

$$
\begin{aligned}
f'(x; u) &= \lim_{\delta \downarrow 0} \frac{|x + \delta u - 1| + (x + \delta u)^2 - |x - 1| - x^2}{\delta} \\
&= \lim_{\delta \downarrow 0} \frac{x + \delta u - 1 + (x + \delta u)^2 - (x - 1) - x^2}{\delta} = (1 + 2x)u;
\end{aligned}
$$

when $x < 1$,

$$
\begin{aligned}
f'(x; u) &= \lim_{\delta \downarrow 0} \frac{|x + \delta u - 1| + (x + \delta u)^2 - |x - 1| - x^2}{\delta} \\
&= \lim_{\delta \downarrow 0} \frac{-(x + \delta u - 1) + (x + \delta u)^2 + (x - 1) - x^2}{\delta} = (-1 + 2x)u,
\end{aligned}
$$

and when $x = 1$,

$$f'(x; u) = \lim_{\delta \downarrow 0} \frac{|\delta u| + (x + \delta u)^2 - x^2}{\delta} = |u| + 2xu = |u| + 2u.$$

Now, let $\bar{x}$ be a local minimum. If $\bar{x} > 1$, we need to have

$$(1 + 2\bar{x})u \geqslant 0, \qquad u = \pm 1.$$

This is impossible. Now, if $\bar{x} = 1$, then we need

$$|u| + 2u \geqslant 0, \qquad u = \pm 1,$$

which is impossible for $u = -1$. Finally, if $\bar{x} < 1$, then we need

$$(-1 + 2\bar{x})u \geqslant 0,$$

which leads to

$$\bar{x} = \frac{1}{2}.$$

This is the only point satisfying the necessary condition and $f(x)$ is coercive. Therefore, we must have

$$\min_{x \in \mathbb{R}} f(x) = f\left(\frac{1}{2}\right) = \frac{3}{4}.$$

The above is illustrative and we could use the method for one variable functions introduced in Section 1 of Chapter 2 to find optimal solutions.

Next, we look at the so-called the *second order necessary condition*.

Proposition 1.3. *Let F be open and $\bar{x} \in F$ be a local optimal solution to Problem* (G). *Suppose $f : F \to \mathbb{R}$ is C^2. Then*

$$f_{xx}(\bar{x}) \geqslant 0. \tag{1.3}$$

Proof. Suppose (1.3) fails. Then by the continuity of $f_{xx}(\cdot)$, there is a $u \in \partial B_1(0)$ such that

$$u^\top f_{xx}(\bar{x} + \lambda u)u < 0, \qquad \forall \lambda \in [0, \lambda_0].$$

Hence, using the second order Taylor expansion, we have (note Theorem 4.12 of Chapter 1).

$$f(\bar{x} + \lambda u) - f(\bar{x}) = \lambda f_x(\bar{x})u + \frac{\lambda^2}{2} u^\top f_{xx}(\bar{x} + \theta \lambda u)u$$
$$= \frac{\lambda^2}{2} u^\top f_{xx}(\bar{x} + \theta \lambda u)u < 0,$$

for some $\theta \in (0, 1)$. This contradicts the optimality of $\bar{x}$. ☐

It turns out that a little strengthening the above condition leads to a *second order sufficient condition* for a local optimal solution.

Proposition 1.4. *Let $F \subseteq \mathbb{R}^n$ be open, and let $f : F \to \mathbb{R}$ be C^2. Suppose*

$$f_x(\bar{x}) = 0, \tag{1.4}$$

and

$$f_{xx}(\bar{x}) > 0. \tag{1.5}$$

Then $\bar{x}$ is a strict local optimal solution to Problem (G).

Proof. By the continuity of $f_{xx}(\cdot)$, we see that there exists a $\delta > 0$ such that

$$u^\top f_{xx}(x)u > 0, \qquad \forall x \in B_\delta(\bar{x}), \ u \in \partial B_1(0).$$

Therefore, by the second order Taylor expansion, we have

$$f(\bar{x} + \lambda u) - f(\bar{x}) = \frac{\lambda^2}{2} u^\top f_{xx}(\bar{x} + \theta \lambda u)u > 0, \qquad \forall \lambda \in (0, \delta], \ u \in \partial B_1(0),$$

for some $\theta \in (0, 1)$. This means that $\bar{x}$ is a strict local minimum. ☐

In the case that F is not open, we need the following notion.

Definition 1.5. Let $F \subseteq \mathbb{R}^n$. We say that the constraint F is *inactive* at $x \in F$ if there exists a $\delta > 0$ such that

$$B_\delta(x) \subseteq F. \tag{1.6}$$

Otherwise, we say that the constraint F is *active* at x.

Clearly, we can similarly prove the following result.

Proposition 1.6. *Let $F \subseteq \mathbb{R}^n$ and $f : F \to \mathbb{R}$. Suppose $\bar{x} \in F$ at which F is inactive.*

(i) *If $\bar{x}$ is a local minimum of Problem* (G), *then*

$$f'(\bar{x}; u) \geqslant 0, \qquad \forall u \in \mathbb{R}^n,$$

provided $f(\cdot)$ is Gâteaux differentiable;

$$\nabla f(\bar{x}) = 0,$$

provided $f(\cdot)$ admits a gradient at $\bar{x}$; and

$$f_{xx}(\bar{x}) \geqslant 0,$$

provided $f(\cdot)$ is C^2 near $\bar{x}$.

(ii) *If $f(\cdot)$ is C^2, and*

$$f_x(\bar{x}) = 0, \qquad f_{xx}(\bar{x}) > 0,$$

then $\bar{x}$ is a strict local optimal solution to Problem (G).

More interestingly, we have the following result.

Proposition 1.7. *Let $F \subseteq \mathbb{R}^n$ be closed, and $f : F \to \mathbb{R}$ be continuously differentiable.*

(i) *If $\bar{x} \in F$ is a minimum of $f(\cdot)$ over F, then*

$$\bar{x} \in F_0 \equiv \left\{ x \in F^\circ \mid f_x(x) = 0 \right\} \bigcup \partial F. \tag{1.7}$$

(ii) *If F is compact, then $f(\cdot)$ attains its minimum over F at some point in F_0 defined by* (1.7).

In the case of $n = 1$, the above result is well-known from standard calculus courses. Practically, this is very useful in finding minima of given functions.

Proof. If $\bar{x} \in \partial F$, we are done. Otherwise, $\bar{x} \in F^\circ$. Then by the local minimality of $\bar{x}$, we must have $f_x(\bar{x}) = 0$. $\qquad\square$

Example 1.8. Let
$$F = \{(x, y) \in \mathbb{R}^2 \mid x^2 + y^2 \leqslant 1\}, \qquad f(x, y) = x^2 - y^2.$$

We calculate
$$f_x(x, y) = 2x, \qquad f_y(x, y) = -2y.$$

Hence, $(0, 0)$ is the only point in F at which $f_x = f_y = 0$. Next, on ∂F, we have $x^2 + y^2 = 1$. Thus, on ∂F, one has
$$f\left(x, \pm\sqrt{1 - x^2}\right) = x^2 - (1 - x^2) = 2x^2 - 1 \equiv g(x), \qquad x \in [-1, 1].$$

For the map $g(\cdot)$, we have
$$\max_{x \in [-1, 1]} g(x) = 1, \qquad \min_{x \in [-1, 1]} g(x) = -1.$$

Since $f(0, 0) = 0$, it is neither maximum nor minimum of $f(\cdot, \cdot)$. Hence,
$$\max_{(x, y) \in F} f(x, y) = 1, \qquad \min_{(x, y) \in F} f(x, y) = -1.$$

The following is an extension of the above example to high dimensions.

Example 1.9. Let
$$F = \left\{(x, y) \mid x \in \mathbb{R}^n, \ y \in \mathbb{R}, \ \|x\|^2 + y^2 \leqslant 1\right\},$$
$$f(x, y) = x^\top Q x - \rho y^2, \qquad (x, y) \in \mathbb{R}^{n+1},$$

where $Q \in \mathbb{S}^n$ is positive definite and $\rho > 0$ is a real number. Since $f(x, y)$ is continuous and F is compact, $f(\cdot, \cdot)$ admits a minimum over F. Now, set
$$f_x(x, y) = 2x^\top Q = 0, \qquad f_y(x, y) = -2\rho y = 0.$$

Then $(0, 0)$ is a critical point of $f(x, y)$, and
$$f(0, 0) = 0.$$

Next, the boundary ∂F can be represented by
$$y = \pm\sqrt{1 - \|x\|^2}, \qquad \|x\| \leqslant 1.$$

Let $g(\cdot)$ be the restriction of $f(\cdot,\cdot)$ on ∂F. Then

$$g(x) = x^\top Q x - \rho\big(1 - \|x\|^2\big), \qquad \|x\| \leqslant 1.$$

If $\|x\| = 1$, then $y = 0$, and

$$f(x,y) = g(x) = x^\top Q x.$$

Hence, if we let $\lambda_0 = \min \sigma(Q) > 0$ (since $Q > 0$)

$$\inf_{\|x\|=1} g(x) = \lambda_0 > 0 = f(0,0).$$

For $\|x\| < 1$, setting

$$0 = g_x(x) = 2x^\top Q + 2\rho x^\top,$$

which leads to

$$Q x = -\rho x.$$

Since $Q > 0$ and $\rho > 0$, the above is true only if $x = 0$, and then $y = \pm 1$,

$$f(0, \pm 1) = -\rho < 0 = f(0,0).$$

Hence, the minimum of $f(\cdot,\cdot)$ is attained at $(0, \pm 1)$ with

$$\min_F f(x,y) = -\rho.$$

Exercises

1. Find local minimizers, local maximizers, global minimizers and global maximizers:

(i) $f(x,y) = x^2 - 4x + 2y^2 - 1$;

(ii) $f(x,y,z) = 3x^2 + 2y^2 + z^2 + 2xy + 4xz + yz$;

(iii) $f(x,y,z) = xyz + \frac{1}{x} + \frac{1}{y} + \frac{1}{z}$, $x, y, z \neq 0$

(iv) $f(x,y) = (x+y)(xy+1)$.

2. Find global maxima and global minima for the given function $f(\cdot)$ over its domain F:

(i) $f(x) = x^3 - 2x^2 - 5x + 6$, $F = [4, 8]$.

(ii) $f(x,y) = x^2 + y^2$, $F = \{(x,y) \in \mathbb{R}^2 \mid x^2 - y^2 \leqslant 1\}$.

(iii) $f(x,y) = x^2 - y^2$, $F = \{(x,y) \in \mathbb{R}^2 \mid x^2 + y^2 \leqslant 1\}$.

3. Construct an example that $f : F \to \mathbb{R}^n$ attains a local minimum at $\bar{x} \in F$ such that for some direction $u \in \partial B_1(0)$, $f'(\bar{x}; u) > 0$.

4. Construct an example that $f : F \to \mathbb{R}$ attains a local minimum at $\bar{x} \in F$, $f(\cdot)$ admits a gradient at $\bar{x}$, but $f(\cdot)$ is not Fréchet differentiable at $\bar{x}$. Check that $\nabla f(\bar{x}) = 0$ (and $f_x(\bar{x}) = 0$ does not make sense).

5. Let $f : F \to \mathbb{R}$ such that $f(\cdot)$ is Fréchet differentiable at $\bar{x}$ and F is inactive at $\bar{x}$. Suppose that for some $\delta > 0$, it holds that

$$f_x(\bar{x} + \lambda u)u \geqslant 0, \qquad \forall u \in \partial B_1(0), \ \lambda \in (0, \delta).$$

Show that $\bar{x}$ is a local optimal solution to Problem (G).

6. Let $f(x) = x^\top A x + 2b^\top x$ for some $A \in \mathbb{R}^{n \times n}$ (not necessarily symmetric) and $b \in \mathbb{R}^n$. Let $\bar{x} \in \mathbb{R}^n$ be a minimum of $f(x)$. Find an expression for $\bar{x}$ and state the condition such that the minimum is unique.

7. Let $f(x, y) = |x + y| + (x - 1)^2 + (y + 1)^2$. Find the minimum of $f(x, y)$.

2   Problems with Equality Constraints

Let $g : \mathbb{R}^n \to \mathbb{R}^m$. Define

$$F = \{x \in \mathbb{R}^n \mid g(x) = 0\}. \tag{2.1}$$

We assume that $F \neq \varnothing$ and $g(\cdot)$ is continuous. Then F is closed. Now, let $f : \mathbb{R}^n \to \mathbb{R}$. Then our Problem (G) becomes

$$\begin{cases} \min \quad f(x), \\ \text{subject to} \quad g(x) = 0. \end{cases} \tag{2.2}$$

We refer to the above problem as an NLP problem with *equality constraints*. We assume that

$$n > m. \tag{2.3}$$

Roughly speaking there are m constraints for n variables, which leave a freedom of degree $(n - m)$. Note that if $n \leqslant m$, then as long as the constraints are not redundant, we might end up with either F is empty or a singleton, which will lead to the problem (2.2) trivial.

The following result is essentially due to Lagrange and it will lead to the so-called *Lagrange multiplier method*.

Theorem 2.1. (i) *Let $\bar{x} \in \mathbb{R}^n$ be a local optimal solution of problem (2.2). Suppose $f(\cdot)$ and $g(\cdot)$ are C^1, and $g_x(\bar{x}) \in \mathbb{R}^{m \times n}$ is of full rank. Then there exists a $\lambda \in \mathbb{R}^m$ such that*

$$f_x(\bar{x}) + \lambda^\top g_x(\bar{x}) = 0. \tag{2.4}$$

If in addition, $f(\cdot)$ and $g(\cdot)$ are C^2 and we define

$$L(x, \lambda) = f(x) + \lambda^\top g(x), \qquad (x, \lambda) \in \mathbb{R}^n \times \mathbb{R}^m, \tag{2.5}$$

then

$$a^\top L_{xx}(\bar{x}, \lambda)a \geqslant 0, \qquad \forall a \in \mathcal{N}\big(g_x(\bar{x})\big). \tag{2.6}$$

(ii) *Suppose* $f(\cdot)$ *and* $g(\cdot)$ *are* C^2 *and for* $\bar{x} \in \mathbb{R}^n$ *with* $g(\bar{x}) = 0$, *and there exists a* $\lambda \in \mathbb{R}^m$ *such that* (2.4) *holds. Moreover,*

$$a^\top L_{xx}(\bar{x}, \lambda)a > 0, \qquad \forall a \in \mathcal{N}\big(g_x(\bar{x})\big), \ a \neq 0, \tag{2.7}$$

then $\bar{x}$ *is a local minimum of Problem* (2.2).

Proof. (i) Take any $a \in \mathcal{N}\big(g_x(\bar{x})\big) \subseteq \mathbb{R}^n$, one has

$$g_x(\bar{x})a = 0. \tag{2.8}$$

Such a vector a is called a *tangent vector* of the set $g(x) = 0$ at $\bar{x}$. We claim that there exists a curve $\varphi(\cdot)$ on the set $g^{-1}(0) \equiv \{x \in \mathbb{R}^n \mid g(x) = 0\}$ passing through $\bar{x}$ with the tangent vector a at $\bar{x}$. To show that, we consider the following equation:

$$\Theta(t, u) \equiv g\big(\bar{x} + ta + g_x(\bar{x})^\top u\big) = 0, \tag{2.9}$$

with $(t, u) \in \mathbb{R} \times \mathbb{R}^m$. Clearly,

$$\Theta(0, 0) = g(\bar{x}) = 0,$$

and

$$\Theta_u(0, 0) = g_x(\bar{x})g_x(\bar{x})^\top \in \mathbb{R}^{m \times m},$$

which is non-singular (since $g_x(\bar{x})$ has rank m). Hence, by implicit function theorem, one can find a function $u : [-\delta, \delta] \to \mathbb{R}^m$ for some $\delta > 0$, such that

$$\begin{cases} u(0) = 0, \\ \Theta(t, u(t)) \equiv g\big(\bar{x} + ta + g_x(\bar{x})^\top u(t)\big) = 0, \qquad t \in [-\delta, \delta]. \end{cases}$$

It is clear that

$$\begin{aligned} 0 = \frac{d}{dt}\Theta(t, u(t)) &= \Theta_t(t, u(t)) + \Theta_u(t, u(t))u'(t) \\ &= g_x(\bar{x} + ta + g_x(\bar{x})^\top u(t))\big[a + g_x(\bar{x})^\top u'(t)\big]. \end{aligned}$$

Evaluating the above at $t = 0$ leads to (noting (2.8))

$$0 = g_x(\bar{x})a + g_x(\bar{x})g_x(\bar{x})^\top u'(0) = g_x(\bar{x})g_x(\bar{x})^\top u'(0).$$

Thus, it is necessary that $u'(0) = 0$. Now, we define

$$\varphi(t) = \bar{x} + ta + g_x(\bar{x})^\top u(t), \qquad t \in [-\delta, \delta].$$

Then one has

$$\begin{cases} g(\varphi(t)) = 0, & t \in [-\delta, \delta], \\ \varphi(0) = \bar{x}, & \varphi'(0) = a. \end{cases} \tag{2.10}$$

Next, by the optimality of $\bar{x}$, we have that $t \mapsto f(\varphi(t))$ attains a local minimum at $t = 0$. Thus,

$$0 = \frac{d}{dt} f(\varphi(t)) \Big|_{t=0} = f_x(\varphi(t)) \varphi'(t) \Big|_{t=0} = f_x(\bar{x}) a. \tag{2.11}$$

Combining (2.8) and (2.11), we have

$$\mathcal{N}\big(g_x(\bar{x})\big) \subseteq \mathcal{N}\big(f_x(\bar{x})\big).$$

This is equivalent to the following: (see Chapter 1, Proposition 2.5)

$$f_x(\bar{x})^\top \in \mathcal{R}\big(g_x(\bar{x})^\top\big),$$

where $\mathcal{R}\big(g_x(\bar{x})^\top\big)$ is the range of $g_x(\bar{x})^\top \in \mathbb{R}^{n \times m}$. Hence, we can find a $\lambda \in \mathbb{R}^m$ such that (2.4) holds.

Next, when $f(\cdot)$ and $g(\cdot)$ are C^2, so is $\varphi(\cdot)$. Consequently, for any $a \in \mathcal{N}(g_x(\bar{x}))$, we have

$$0 \leqslant \frac{d^2}{dt^2} f(\varphi(t)) \Big|_{t=0} = \left[\varphi'(t)^\top f_{xx}(\varphi(t)) \varphi'(t) + f_x(\varphi(t)) \varphi''(t) \right] \Big|_{t=0} \tag{2.12}$$

$$= a^\top f_{xx}(\bar{x}) a + f_x(\bar{x}) \varphi''(0).$$

On the other hand, letting $\theta(x, \lambda) = \lambda^\top g(x)$, we have

$$\theta(\varphi(t), \lambda) = g(\varphi(t))^\top \lambda = 0, \quad t \in [-\delta, \delta].$$

Differentiating the above twice with respect to t, we have

$$0 = \varphi'(t)^\top \theta_{xx}(\varphi(t), \lambda) \varphi'(t) + \theta_x(\varphi(t); \lambda) \varphi''(t)$$

$$= \varphi'(t)^\top \theta_{xx}(\varphi(t), \lambda) \varphi'(t) + \lambda^\top g_x(\varphi(t)) \varphi''(t).$$

Hence, evaluating at $t = 0$, we have

$$a^\top \theta_{xx}(\bar{x}, \lambda) a + \lambda^\top g_x(\bar{x}) \varphi''(0) = 0.$$

Consequently, by (2.4) and (2.12),

$$0 \leqslant a^\top f_{xx}(\bar{x}) a + f_x(\bar{x}) \varphi''(0) = a^\top f_{xx}(\bar{x}) a - \lambda^\top g_x(\bar{x}) \varphi''(0)$$

$$= a^\top f_{xx}(\bar{x}) a + a^\top \theta_{xx}(\bar{x}, \lambda) a$$

$$= a^\top \left[f_{xx}(\bar{x}) + \big(\lambda^\top g(x)\big)_{xx} \Big|_{x=\bar{x}} \right] a = a^\top L_{xx}(\bar{x}, \lambda) a.$$

This prove (2.6).

(ii) Suppose $\bar{x}$ is not a local minimum of problem (2.2). Then we can find a sequence s_k, $\|s_k\| = 1$, and $\delta_k > 0$, $\delta_k \to 0$ such that $\bar{x} + \delta_k s_k$ is feasible for each $k \geqslant 1$ and

$$f(\bar{x} + \delta_k s_k) < f(\bar{x}), \qquad k \geqslant 1. \tag{2.13}$$

We may assume that $s_k \to \bar{s}$, with $\|\bar{s}\| = 1$. Clearly, (2.13) implies

$$0 > f(\bar{x} + \delta_k s_k) - f(\bar{x}) = \delta_k f_x(\bar{x}) s_k + \frac{\delta_k^2}{2} s_k^\top f_{xx}(\bar{x}) s_k + o(\delta_k^2).$$

On the other hand, since $\bar{x} + \delta_k s_k$ is feasible,

$$0 = g^i(\bar{x} + \delta_k s_k) = g^i(\bar{x}) + \delta_k g_x^i(\bar{x}) s_k + \frac{\delta_k^2}{2} s_k^\top g_{xx}^i(\bar{x}) s_k + o(\delta_k^2)$$

$$= \delta_k g_x^i(\bar{x}) s_k + \frac{\delta_k^2}{2} s_k^\top g_{xx}^i(\bar{x}) s_k + o(\delta_k^2).$$

Therefore, we must have

$$g_x^i(\bar{x}) \bar{s} = 0, \qquad 1 \leqslant i \leqslant m,$$

which means that $\bar{s} \in \mathcal{N}(g_x(\bar{x})) \setminus \{0\}$. Then, it follows that

$$0 > \delta_k \big[f_x(\bar{x}) + \lambda^\top g_x(\bar{x}) \big] s_k + \frac{\delta_k^2}{2} s_k^\top \Big[f_{xx}(\bar{x}) + \sum_{i=1}^m \lambda_i g_{xx}^i(\bar{x}) \Big] s_k + o(\delta_k^2)$$

$$= \frac{\delta_k^2}{2} s_k^\top L_{xx}(\bar{x}, \lambda) s_k + o(\delta_k^2).$$

Sending $k \to \infty$, one has

$$\bar{s}^\top L_{xx}(\bar{x}, \lambda) \bar{s} \leqslant 0.$$

This contradicts the positive definiteness of $L_{xx}(\bar{x}, \lambda)$ on $\mathcal{N}(g_x(\bar{x}))$, proving the sufficiency. $\qquad\square$

The vector λ in the above is called a *Lagrange multiplier*. The condition that $g_x(\bar{x})$ is of full rank is called the *regularity* of the constraint $g(\cdot)$ at $\bar{x}$. Note that $g_x(\bar{x}) \in \mathbb{R}^{m \times n}$ and we have assumed $n > m$. Thus, $g_x(\bar{x})$ is of full rank means that

$$\text{rank}\,(g_x(\bar{x})) = m,$$

which is also equivalent to the linear independence of the vectors $g_x^1(\bar{x})^\top$, $g_x^2(\bar{x})^\top, \cdots, g_x^m(\bar{x})^\top$.

Function $L(x, \lambda)$ defined by (2.5) is called the *Lagrangian* or *Lagrange function* of problem (2.2). Condition (2.4) can be written as

$$L_x(\bar{x}, \lambda) = 0, \tag{2.14}$$

and the equality constraint is equivalent to

$$L_\lambda(\bar{x}, \lambda) = g(\bar{x}) = 0. \tag{2.15}$$

Hence, the point $(\bar{x}, \lambda)$ satisfying (2.4) and the equality constraint is nothing but a *critical point* of the function $(x, \lambda) \mapsto L(x, \lambda)$ over the whole space $\mathbb{R}^n \times \mathbb{R}$ (without constraints). On the other hand, with $\theta(x, \lambda) = \lambda^\top g(x)$, one has

$$\begin{aligned} D^2_{(x,\lambda)} L(x, \lambda) &\equiv \begin{pmatrix} L_{xx}(x, \lambda) & L_{\lambda x}(x, \lambda) \\ L_{x\lambda}(x, \lambda) & L_{\lambda\lambda}(x, \lambda) \end{pmatrix} \\ &= \begin{pmatrix} f_{xx}(x, \lambda) + \theta_{xx}(x, \lambda) & g_x(x)^\top \\ g_x(x) & 0 \end{pmatrix}. \end{aligned} \tag{2.16}$$

Thus, for any $a \in \mathbb{R}^n$ and $b \in \mathbb{R}^m$,

$$\begin{pmatrix} a \\ b \end{pmatrix}^\top D^2_{(x,\lambda)} L(x, \lambda) \begin{pmatrix} a \\ b \end{pmatrix} = a^\top [f_{xx}(x) + \theta_{xx}(x, \lambda)] a + 2b^\top g_x(x) a. \tag{2.17}$$

Hence, (2.6) is equivalent to the positive semi-definiteness of $D^2_{(x,\lambda)} L(\bar{x}, \lambda)$ on $\mathcal{N}(g_x(\bar{x})) \times \mathbb{R}^m$ only, not on the whole $\mathbb{R}^n \times \mathbb{R}^m$. Therefore, the critical point $(\bar{x}, \lambda)$ of $L(x, \lambda)$ might not be a minimum of $L(x, \lambda)$. Hence, to check if $\bar{x}$ is a minimum of $f(\cdot)$ over the feasible set F, one has to use (2.7). All these suggest that one can solve problem (2.2) by finding critical points of the function $L(x, \lambda)$ over the whole space $\mathbb{R}^n \times \mathbb{R}^m$. Then use (2.7) to test if $\bar{x}$ is a local minimum. This method is referred to as the *Lagrange multiplier method*.

It is not surprising that we have a similar result for the following problem:

$$\begin{cases} \max \quad f(x), \\ \text{subject to} \quad g(x) = 0. \end{cases} \tag{2.18}$$

More precisely, we state the following.

Theorem 2.2. (i) *Let $\bar{x} \in \mathbb{R}^n$ be a local optimal solution of problem (2.18). Suppose $f(\cdot)$ and $g(\cdot)$ are C^1, and $g_x(\bar{x}) \in \mathbb{R}^{m \times n}$ is of full rank. Then there exists a $\lambda \in \mathbb{R}^m$ such that (2.4) holds. If in addition, $f(\cdot)$ and $g(\cdot)$ are C^2 and we define $L(x, \lambda)$ as (2.5), then*

$$a^\top L_{xx}(\bar{x}, \lambda) a \leqslant 0, \qquad \forall a \in \mathcal{N}(g_x(\bar{x})). \tag{2.19}$$

(ii) *Suppose $f(\cdot)$ and $g(\cdot)$ are C^2 and for $\bar{x} \in \mathbb{R}^n$ with $g(\bar{x}) = 0$ and there exists a $\lambda \in \mathbb{R}^m$ such that (2.4) holds. Moreover,*

$$a^\top L_{xx}(\bar{x}, \lambda)a < 0, \qquad \forall a \in \mathcal{N}\Big(g_x(\bar{x})\Big), \ a \neq 0, \qquad (2.20)$$

then $\bar{x}$ is a local maximum of Problem (2.18).

We leave the proof to the readers. Now, let us present an example showing how one can use Lagrange multiplier method to solve a problem of form (2.2).

Example 2.3. Let

$$f(x, y) = -(x - 1)^2 - (y - 2)^2, \qquad g(x, y) = x^2 + y^2 - 4.$$

We want to find the minimum of $f(x, y)$ subject to $g(x, y) = 0$. To this end, we define Lagrangian

$$L(x, y, \lambda) = -(x - 1)^2 - (y - 2)^2 + \lambda(x^2 + y^2 - 4).$$

Set

$$0 = L_x(x, y, \lambda) = -2(x - 1) + 2\lambda x = 2\Big[(\lambda - 1)x + 1\Big],$$

$$0 = L_y(x, y, \lambda) = -2(y - 2) + 2\lambda y = 2\Big[(\lambda - 1)y + 2\Big].$$

Thus, $\lambda \neq 1$, and

$$x = \frac{1}{1 - \lambda}, \qquad y = \frac{2}{1 - \lambda}.$$

Then by the constraint, one has

$$4 = x^2 + y^2 = \frac{1}{(1 - \lambda)^2} + \frac{4}{(1 - \lambda)^2} = \frac{5}{(1 - \lambda)^2},$$

which leads to

$$\lambda = 1 \pm \frac{\sqrt{5}}{2}.$$

From this, we obtain two critical points of $L(x, y, \lambda)$:

$$(x_1, y_1, \lambda_1) = \Big(\frac{2}{\sqrt{5}}, \frac{4}{\sqrt{5}}, 1 - \frac{\sqrt{5}}{2}\Big),$$

$$(x_2, y_2, \lambda_2) = \Big(-\frac{2}{\sqrt{5}}, -\frac{4}{\sqrt{5}}, 1 + \frac{\sqrt{5}}{2}\Big).$$

Since

$$\nabla g(x, y) = (2x, 2y),$$

one sees that

$$\nabla g(x_1, y_1) = \frac{2}{\sqrt{5}}(1, 2), \qquad \nabla g(x_2, y_2) = -\frac{2}{\sqrt{5}}(1, 2).$$

Both are of full rank. Hence, the constraint $g(x, y) = 0$ is regular at the critical points. On the other hand,

$$\mathcal{N}\left(\nabla g(x_i, y_i)\right) = \left\{a \begin{pmatrix} 2 \\ -1 \end{pmatrix} \mid a \in \mathbb{R}\right\}, \qquad i = 1, 2.$$

Next, we calculate

$$D^2_{(x,y)} L(x, y, \lambda) = \begin{pmatrix} -2 + 2\lambda & 0 \\ 0 & -2 + 2\lambda \end{pmatrix} = 2(\lambda - 1)I = \pm\sqrt{5}\,I.$$

Hence,

$$\begin{pmatrix} 2 \\ -1 \end{pmatrix}^{\top} D^2_{(x,y)} L\left(\frac{2}{\sqrt{5}}, \frac{4}{\sqrt{5}}, 1 - \frac{\sqrt{5}}{2}\right) \begin{pmatrix} 2 \\ -1 \end{pmatrix} = -5\sqrt{5} < 0,$$

and

$$\begin{pmatrix} 2 \\ -1 \end{pmatrix}^{\top} D^2_{(x,y)} L\left(-\frac{2}{\sqrt{5}}, -\frac{4}{\sqrt{5}}, 1 + \frac{\sqrt{5}}{2}\right) \begin{pmatrix} 2 \\ -1 \end{pmatrix} = 5\sqrt{5} > 0.$$

This leads to the following conclusion: $(\frac{2}{\sqrt{5}}, \frac{4}{\sqrt{5}})$ is the local maximum and $(-\frac{2}{\sqrt{5}}, -\frac{4}{\sqrt{5}})$ is the local minimum.

It seems to be a little complicated by using the second order sufficient condition to check the optimality of the Lagrangian's critical points. However, sometimes it is possible to avoid using the second order conditions. Actually, if $f(\cdot)$ and $g(\cdot)$ are continuous and the set $\{x \in \mathbb{R}^n \mid g(x) = 0\}$ is bounded, then it must be compact, and Weierstrass Theorem will be applicable. If we let

$$\mathcal{X} = \{x \in \mathbb{R}^n \mid (x, \lambda) \text{ is a critical point of } L(x, \lambda) \text{ for some } \lambda\},$$

then all the global maximum and minimum of $f(\cdot)$ over $g^{-1}(0)$ must be in $\mathcal{X}$. Hence, we need only find maximum and minimum of $f(\cdot)$ over $\mathcal{X}$. Practically, if the set $\mathcal{X}$ is finite, we may just evaluate $f(\cdot)$ over $\mathcal{X}$ to get maxima and minima. Let us present an example to illustrate this.

Example 2.4. Let $f(x, y) = xy$ and $g(x, y) = x^2 + (y + 1)^2 - 1$. We want to find maximum and minimum of $f(\cdot, \cdot)$ over $g^{-1}(0)$. Note that $g^{-1}(0)$ is a circle centered at $(0, -1)$ with radius 1. Hence, it is compact. Now, we form

$$L(x, y, \lambda) = xy + \lambda[x^2 + (y + 1)^2 - 1].$$

Set

$$0 = L_x(x, y, \lambda) = y + 2\lambda x, \qquad 0 = L_y(x, y, \lambda) = x + 2\lambda(y + 1).$$

Thus,

$$x = \frac{2\lambda}{4\lambda^2 - 1}, \qquad y = \frac{-4\lambda^2}{4\lambda^2 - 1}.$$

By the constraint, we have

$$1 = x^2 + (y + 1)^2 = \frac{4\lambda^2 + 1}{(4\lambda^2 - 1)^2},$$

which leads to

$$0 = (4\lambda^2 - 1)^2 - 4\lambda^2 - 1 = 16\lambda^4 - 12\lambda^2 = 4\lambda^2(4\lambda^2 - 3).$$

Hence,

$$\lambda = 0, \ \pm \frac{\sqrt{3}}{2}.$$

Then we obtain 3 critical points of $L(x, y, \lambda)$ as follows:

$$(0, 0, 0), \quad \left(\frac{\sqrt{3}}{2}, \frac{\sqrt{3}}{2}, -\frac{3}{2}\right), \quad \left(-\frac{\sqrt{3}}{2}, -\frac{\sqrt{3}}{2}, -\frac{3}{2}\right).$$

This leads to

$$\mathcal{X} = \left\{(0, 0), \left(\frac{\sqrt{3}}{2}, -\frac{3}{2}\right), \left(-\frac{\sqrt{3}}{2}, -\frac{3}{2}\right)\right\}.$$

We now evaluate

$$f(0, 0) = 0, \quad f\left(\frac{\sqrt{3}}{2}, -\frac{3}{2}\right) = -\frac{3\sqrt{3}}{4}, \quad f\left(-\frac{\sqrt{3}}{2}, -\frac{3}{2}\right) = \frac{3\sqrt{3}}{4}.$$

Therefore,

$$\max_{g(x,y)=0} f(x, y) = f\left(-\frac{\sqrt{3}}{2}, -\frac{3}{2}\right) = \frac{3\sqrt{3}}{4},$$

$$\min_{g(x,y)=0} f(x, y) = f\left(\frac{\sqrt{3}}{2}, -\frac{3}{2}\right) = -\frac{3\sqrt{3}}{4}.$$

Let us apply the second order sufficient conditions. To this end, we calculate

$$\nabla g(x, y) = (2x, 2(y + 1)).$$

Thus,

$$\nabla g\left(\frac{\sqrt{3}}{2}, -\frac{3}{2}\right) = (\sqrt{3}, -1), \quad \nabla g\left(-\frac{\sqrt{3}}{2}, -\frac{3}{2}\right) = (-\sqrt{3}, -1).$$

Consequently,

$$\mathcal{N}\left(\nabla g(\tfrac{\sqrt{3}}{2}, -\tfrac{3}{2})\right) = \left\{a\begin{pmatrix}1\\\sqrt{3}\end{pmatrix} \mid a \in \mathbb{R}\right\},$$

$$\mathcal{N}\left(\nabla g(-\tfrac{\sqrt{3}}{2}, -\tfrac{3}{2})\right) = \left\{a\begin{pmatrix}1\\-\sqrt{3}\end{pmatrix} \mid a \in \mathbb{R}\right\}.$$

On the other hand,

$$D^2_{(x,y)}L\left(x, y, \tfrac{\sqrt{3}}{2}\right) = \begin{pmatrix}\sqrt{3} & 1\\1 & \sqrt{3}\end{pmatrix},$$

$$D^2_{(x,y)}L\left(x, y, -\tfrac{\sqrt{3}}{2}\right) = \begin{pmatrix}-\sqrt{3} & 1\\1 & -\sqrt{3}\end{pmatrix}.$$

Hence, on $\mathcal{N}(\nabla g(\tfrac{\sqrt{3}}{2}, -\tfrac{3}{2}))$, we calculate

$$(1, \sqrt{3})\begin{pmatrix}\sqrt{3} & 1\\1 & \sqrt{3}\end{pmatrix}\begin{pmatrix}1\\\sqrt{3}\end{pmatrix} = 6\sqrt{3} > 0,$$

and on $\mathcal{N}(\nabla g(-\tfrac{\sqrt{3}}{2}, -\tfrac{3}{2}))$, we have

$$(1, -\sqrt{3})\begin{pmatrix}-\sqrt{3} & 1\\1 & -\sqrt{3}\end{pmatrix}\begin{pmatrix}1\\-\sqrt{3}\end{pmatrix} = -6\sqrt{3} < 0.$$

Therefore, $(\tfrac{\sqrt{3}}{2}, -\tfrac{3}{2})$ is a local minimum point and $(-\tfrac{\sqrt{3}}{3}, -\tfrac{3}{2})$ is a local maximum point.

Also, we have

$$\nabla g(0,0) = (0, 2).$$

Thus,

$$\mathcal{N}(\nabla g(0,0)) = \left\{\begin{pmatrix}a\\0\end{pmatrix} \mid a \in \mathbb{R}\right\}.$$

Since

$$D^2_{(x,y)}L(0,0,0) = \begin{pmatrix}0 & 1\\1 & 0\end{pmatrix},$$

we have

$$(1, 0)\begin{pmatrix}0 & 1\\1 & 0\end{pmatrix}\begin{pmatrix}1\\0\end{pmatrix} = 0.$$

Therefore, no conclusion can be drawn from the Lagrange multiplier method whether $(0, 0)$ is a local extreme.

From the above, we see the following:

• If F is compact, the global minima/maxima can be completely determined (in principle).

• If F is non-compact, one can only obtain local minima/maxima. It could also have the situation that the test fails.

Next, we look at the following example which will lead to some new issue.

Example 2.5. Consider $f(x,y) = -x^3 - y^3$ and $g(x,y) = x + y - 1$. We try to find minimum of $f(x,y)$ subject to $g(x,y) = 0$. To begin the Lagrange multiplier method, we form the Lagrangian

$$L(x,y,\lambda) = -x^3 - y^3 + \lambda(x + y - 1).$$

Set

$$0 = L_x(x,y,\lambda) = -3x^2 + \lambda, \qquad 0 = L_y(x,y,\lambda) = -3y^2 + \lambda.$$

Then

$$x^2 = y^2 = \frac{\lambda}{3} \triangleq \mu^2,$$

for some $\mu \geqslant 0$. It is necessary that

$$x = \pm\mu, \qquad y = \pm\mu.$$

By the constraint, we see that one must have

$$x = y = \frac{1}{2} = \mu, \qquad \lambda = 3\mu^2 = \frac{3}{4}.$$

Therefore, $(\frac{1}{2}, \frac{1}{2}, \frac{3}{4})$ is the only critical point of $L(x,y,\lambda)$. Note that

$$\nabla g(x,y) = (1,1),$$

which is of full rank everywhere. Hence, the constraint is regular at any point $(x,y) \in \mathbb{R}^2$. Further,

$$\mathcal{N}\left(\nabla g(\tfrac{1}{2}, \tfrac{1}{2})\right) = \left\{ \begin{pmatrix} a \\ b \end{pmatrix} \mid a + b = 0 \right\} = \left\{ a \begin{pmatrix} 1 \\ -1 \end{pmatrix} \mid a \in \mathbb{R} \right\}.$$

Also,

$$D^2_{(x,y)}L(x,y,\lambda) = \begin{pmatrix} -6x & 0 \\ 0 & -6y \end{pmatrix}.$$

Hence,

$$a^2(1,-1)D^2_{(x,y)}L\left(\tfrac{1}{2}, \tfrac{1}{2}, \lambda\right)\begin{pmatrix} 1 \\ -1 \end{pmatrix}$$

$$= a^2(1,-1)\begin{pmatrix} -3 & 0 \\ 0 & -3 \end{pmatrix}\begin{pmatrix} 1 \\ -1 \end{pmatrix} = -6a^2 < 0.$$

Thus, Lagrange multiplier method gives a conclusion that $(\frac{1}{2}, \frac{1}{2})$ is a local maximum, and we fail to find a local minimum.

As a matter of fact, in the above example, we actually have

$$\inf_{g(x,y)=0} f(x,y) = -\infty.$$

To see this, we need only to make substitution $y = 1 - x$ into $f(x,y)$ to get

$$\widetilde{f}(x) = f(x, 1-x) = -x^3 - (1-x)^3 = -3x^2 + 3x - 1,$$

whose infimum is $-\infty$. Note that in this example, the corresponding feasible set $F = \{(x,y) \in \mathbb{R}^2 \mid g(x,y) = 0\}$ is unbounded and $f(x,y)$ is not coercive on F. Thus, the existence of an optimal solution to the corresponding NLP problem is not guaranteed. Moreover, setting

$$0 = \widetilde{f}'(x) = -6x + 3,$$

we get $x = \frac{1}{2}$ and

$$\widetilde{f}''(\frac{1}{2}) = -6 < 0.$$

Thus, $\widetilde{f}(\cdot)$ does not have a local minimum. Consequently, $f(\cdot,\cdot)$ does not have a local minimum. The above example shows that if no local optimal solutions exist, Lagrange multiplier method might fail.

We now present another example which will lead to another subtle situation.

Example 2.6. Let $f(x,y) = x$ and $g(x,y) = x^3 - y^2$. We want to find (local) minimum and maximum of $f(x,y)$ subject to $g(x,y) = 0$. To this end, we form the following Lagrangian:

$$L(x,y,\lambda) = x + \lambda(x^3 - y^2).$$

Then set

$$0 = L_x(x,y,\lambda) = 1 + 3\lambda x^2, \qquad 0 = L_y(x,y,\lambda) = -2\lambda y.$$

Thus, either $\lambda = 0$ or $y = 0$. But, from the first equation in the above, $\lambda \neq 0$. Hence, $y = 0$ is the only choice. Then the constraint gives $x = 0$, which again contradicts the first equation. Hence, we end up with nothing from the Lagrange multiplier method.

Let us look at the reason causing this. From the constraint, we have

$$x^3 = y^2 \geqslant 0.$$

Hence, $x \in [0, \infty)$. Then we see that

$$\min_{g(x,y)=0} f(x,y) = \min_{x \geqslant 0} x = f(0,0) = 0.$$

It is clear that

$$\nabla g(0,0) = (0,0),$$

which is not of full rank. This means that the constraint $g(x,y) = 0$ is not regular at the optimal solution $(0,0)$. We also see that

$$\nabla f(0,0) = (1,0).$$

Hence, there is no $\lambda \in \mathbb{R}$ such that

$$\nabla f(0,0) + \lambda \nabla g(0,0) = 0.$$

From here, we see how important the regularity of the constraint at the optimal solution is.

We briefly summarize the above two examples as follows: Lagrange multiplier method might not work if either the local optimal solution does not exist, or the constraint is not regular at the local optimal solution.

Now, let us look at optimal solutions to some interesting special cases. We first consider the following problem:

$$\begin{cases} \min & x^\top Q x + 2c^\top x, \\ \text{subject to} & Ax = b, \end{cases} \tag{2.21}$$

where $Q \in \mathbb{S}^n$, $c \in \mathbb{R}^n$, $A \in \mathbb{R}^{m \times n}$, and $b \in \mathbb{R}^m$. We assume that

$$\text{rank}\, A = \text{rank}\, (A, b) = m < n. \tag{2.22}$$

Then the feasible set

$$F \equiv \{x \in \mathbb{R}^n \mid Ax = b\}$$

contains infinitely many points. We have the following result.

Proposition 2.7. *Suppose that $Q \in \mathbb{S}^n$ is invertible and (2.22) holds. Further, suppose Q is positive definite on $\mathcal{N}(A)$. Then problem (2.21) admits a unique optimal solution given by*

$$\bar{x} = [Q^{-1}A^\top(AQ^{-1}A^\top)^{-1}AQ^{-1} - Q^{-1}]c + Q^{-1}A^\top(AQ^{-1}A^\top)^{-1}b, \tag{2.23}$$

with the minimum value of the objective function:

$$(b + AQ^{-1}c)^\top(AQ^{-1}A^\top)^{-1}(b + AQ^{-1}c) - c^\top Q^{-1}c. \tag{2.24}$$

Proof. Form the following Lagrange function

$$L(x, \lambda) = x^\top Q x + 2c^\top x + \lambda^\top (Ax - b).$$

Set

$$0 = L_x(x, \lambda) = 2x^\top Q + 2c^\top + \lambda^\top A.$$

Thus, we have the unique solution $\bar{x}$ of the above:

$$\bar{x} = -Q^{-1}\left(\frac{1}{2}A^\top \lambda + c\right). \tag{2.25}$$

Now, by the equality constraint in (2.21), we obtain

$$b = A\bar{x} = -AQ^{-1}\left(\frac{1}{2}A^\top \lambda + c\right) = -\frac{1}{2}(AQ^{-1}A^\top)\lambda - AQ^{-1}c.$$

Since A has rank m, so is $(AQ^{-1}A^\top)$, which leads to the invertibility of $(AQ^{-1}A^\top)$. Then

$$\lambda = -2(AQ^{-1}A^\top)^{-1}(AQ^{-1}c + b).$$

Substituting the above into (2.25), we obtain (2.23). Now, we look at the second order condition. Note that

$$L_{xx}(x, \lambda) = Q, \qquad \mathcal{N}(g_x(x)) = \mathcal{N}(A),$$

and by our condition, $L_{xx}(x, \lambda)$ is positive definite on $\mathcal{N}(A)$. Hence, $\bar{x}$ given by (2.23) is a minimum. Since such a representation is unique, we obtain the uniqueness of the optimal solution. Finally, we calculate the minimum value of the objective function.

$$
\begin{aligned}
\bar{x}^\top Q\bar{x} + 2c^\top \bar{x} &= -\bar{x}^\top\left(\frac{1}{2}A^\top \lambda + c\right) + 2c^\top \bar{x} \\
&= -\frac{1}{2}\lambda^\top A\bar{x} + c^\top \bar{x} = -\frac{1}{2}\lambda^\top b - c^\top Q^{-1}\left(\frac{1}{2}A^\top \lambda + c\right) \\
&= -\frac{1}{2}\lambda^\top(b + AQ^{-1}c) - c^\top Q^{-1}c \\
&= (b + AQ^{-1}c)^\top(AQ^{-1}A^\top)^{-1}(b + AQ^{-1}c) - c^\top Q^{-1}c,
\end{aligned}
\tag{2.26}
$$

giving the minimum value (2.24) of the objective function. $\qquad\square$

Next, we consider a further specified case:

$$
\begin{cases}
\min & c^\top x, \\
\text{subject to} & Ax = b,
\end{cases}
\tag{2.27}
$$

where $A \in \mathbb{R}^{m \times n}$, $c \in \mathbb{R}^n$, and $b \in \mathbb{R}^m$ are given. We have the following result.

Proposition 2.8. *Suppose* (2.22) *holds. Then Problem* (2.27) *admits a finite minimum value of objective function if and only if*

$$c \in \mathcal{R}(A^\top). \tag{2.28}$$

In this case, any $x \in F$ is optimal and the minimum value of the objective function is given by

$$c^\top A^\top (AA^\top)^{-1} b. \tag{2.29}$$

Proof. Let $x \in F \equiv \{x \in \mathbb{R}^n \mid Ax = b\}$ be fixed. For any $y \in \mathcal{N}(A)$, we have $x + y \in F$. Thus, in order the objective function to have a finite infimum, it is necessary that

$$c^\top y = 0, \qquad \forall y \in \mathcal{N}(A).$$

This means that

$$\mathcal{N}(A) \subseteq \mathcal{N}(c^\top),$$

which is equivalent to (2.28). From (2.28), we can find a $\lambda \in \mathbb{R}^m$ such that

$$c = A^\top \lambda.$$

Due to (2.22), we have

$$\lambda = (AA^\top)^{-1} Ac,$$

which is unique. Hence,

$$c^\top x = (A^\top \lambda)^\top x = \lambda^\top Ax = \lambda^\top b = c^\top A^\top (AA^\top)^{-1} b, \qquad \forall x \in F.$$

Thus, any $x \in F$ is optimal with the above minimum value of the objective function. $\qquad \square$

Exercises

1. Let $f(x, y) = x^2 - y^2$ and $g(x, y) = x^2 + y^2 - 1$. Find the maximum and minimum of $f(x, y)$ subject to $g(x, y) = 0$ by Lagrange multiplier method. Also, using substitution $y^2 = 1 - x^2$ to transform the problem into an optimization of a single variable function, and find the corresponding maximum and minimum. Compare the results.

2. Let $f(x, y) = x^3 + y^3$ and $g(x, y) = x + y - 1$. Use Lagrange multiplier method to find critical points of the Lagrangian. Is this critical point a local maximum or minimum of $f(x, y)$ subject to $g(x, y) = 0$? Why?

3. Let $f(x) = \|x\|^2$ and $g(x) = Ax - b$ with $A \in \mathbb{R}^{m \times n}$ and $b \in \mathbb{R}^m$. Find a minimum of $f(x)$ subject to $g(x) = 0$ by Lagrange multiplier method.

4. Let $f(x, y) = x^4 + y^2$ and $g(x, y) = (x - 2)^3 - 2y^2$. Find minimum of $f(x, y)$ subject to $g(x, y) = 0$ by Lagrange multiplier method, and by a direct method. Explain your results.

5. Let $f(x, y, z) = x + y + z$ and $g(x, y, z) = \frac{1}{x} + \frac{1}{y} + \frac{1}{z} - 1$. find maximum and minimum of $f(x, y, z)$ subject to $g(x, y, z) = 0$.

6. Let $f(x, y, z) = xyz$ and $g(x, y, z) = \begin{pmatrix} x + y + z - 5 \\ xy + yz + zx - 8 \end{pmatrix}$. Find maxima and minima of $f(x, y, z)$ subject to $g(x, y, z) = 0$.

3 Problems with Equality and Inequality Constraints

We now look at problems with equality and inequality constraints. More precisely, let $f : G \to \mathbb{R}$, $g : G \to \mathbb{R}^m$, and $h : G \to \mathbb{R}^\ell$, with $G \subseteq \mathbb{R}^n$ being a domain. Our optimization problem can be stated as follows.

$$\begin{cases} \min \quad f(x), \\ \text{subject to} \quad g(x) = 0, \quad h(x) \leqslant 0. \end{cases} \tag{3.1}$$

This problem is referred to as an NLP problem with equality and *inequality constraints*. In this case, the feasible set takes the following form:

$$F = \{x \in G \mid g(x) = 0, \ h(x) \leqslant 0\}.$$

We still assume that $n > m$. As we pointed out in the previous section that such a condition is almost necessary for the corresponding NLP problem to be non-trivial.

Before presenting some general results, let us look at a couple of examples.

Example 3.1. Let $f(x, y) = x^2 + y^2$, $g(x, y) = 2x^2 + y^2 - 1$, $h(x, y) = x + y - 1$. We want to find the minimum and maximum of $f(x, y)$ subject to $g(x, y) = 0$ and $h(x, y) \leqslant 0$. By the equality constraint, one has $1 - 2x^2 = y^2 \geqslant 0$. Thus, the feasible range for x is $0 \leqslant |x| \leqslant \frac{\sqrt{2}}{2}$, and with $y = \pm\sqrt{1 - 2x^2}$, the objective function becomes

$$f(x, \pm\sqrt{1 - 2x^2}) = 1 - x^2.$$

Hence,

$$\max_{g(x,y)=0} f(x, y) = f(0, \pm 1) = 1,$$

and

$$\min_{g(x,y)=0} f(x,y) = f\left(\pm\frac{\sqrt{2}}{2},0\right) = \frac{1}{2}.$$

We also see that

$$h(0,\pm1) = \pm1 - 1 \leqslant 0, \qquad h\left(\pm\frac{\sqrt{2}}{2},0\right) = \pm\frac{\sqrt{2}}{2} - 1 < 0.$$

Therefore, for this example, the original problem with equality and inequality constraints is equivalent to the problem with equality constraint only. In another word, for this problem, we may solve the problem with equality constraint only and find that the solutions to such a problem automatically satisfy the inequality constraint. Therefore the solutions to the problem with equality constraint only also give the solutions to the problem with equality and inequality constraints.

We now look at the following example, which is a small modification of the above one.

Example 3.2. Let $f(x,y)$ and $g(x,y)$ be the same as above, and let $h(x,y) = xy + \frac{\sqrt{2}}{4}$. We again would like to find maximum and minimum of $f(x,y)$ subject to $g(x,y) = 0$ and $h(x,y) \leqslant 0$. Now, if we still use the same way as the above example, we obtain the minimum points $(\pm\frac{\sqrt{2}}{2},0)$ and maximum points $(0,\pm1)$ for the problem with the equality constraint $g(x,y) = 0$ only. Clearly, all these points do not satisfy the inequality constraint $h(x,y) \leqslant 0$, since

$$h(\pm\frac{\sqrt{2}}{2},0) = \frac{\sqrt{2}}{4} > 0, \qquad h(0,\pm1) = \frac{\sqrt{2}}{4} > 0.$$

Thus, we cannot use the above method. As a matter of fact, in the current example, from $g(x,y) = 0$, we have

$$y = \pm\sqrt{1-2x^2}, \qquad 0 \leqslant |x| \leqslant \frac{\sqrt{2}}{2}.$$

Then (x,y) is feasible if and only if

$$y = \pm\sqrt{1-2x^2}, \qquad \pm x\sqrt{1-2x^2} + \frac{\sqrt{2}}{4} \leqslant 0,$$

which implies

$$\frac{\sqrt{2}}{4} \leqslant |x|\sqrt{1-2|x|^2}.$$

Then

$$\frac{1}{8} \leqslant |x|^2(1 - 2|x|^2) \equiv \varphi(|x|).$$

Set

$$0 = \varphi'(r) = 2r - 8r^3 = 2r(1 - 4r^2),$$

which leads to $r = \frac{1}{2}$, a maximum of $\varphi(r)$, with

$$\varphi(\frac{1}{2}) = \frac{1}{8}.$$

Hence, the only feasible points are $\pm(\frac{1}{2}, -\frac{\sqrt{2}}{2})$. Consequently,

$$\max_{g(x,y)=0, h(x,y)\leqslant 0} f(x,y) = \min_{g(x,y)=0, h(x,y)\leqslant 0} f(x,y) = \frac{3}{4}.$$

The above two examples show that when some inequality constraints appear, one might not simply ignore their presence. We now present the following result.

Theorem 3.3. *Let $f(\cdot)$, $g(\cdot)$, and $h(\cdot)$ be C^1.*

(i) *Let $\bar{x}$ be a local optimal solution to Problem (3.1) and the constraints be regular at $\bar{x}$, namely, if*

$$\begin{cases} J(\bar{x}) = \{j \mid h^j(\bar{x}) = 0, \ 1 \leqslant j \leqslant \ell\} \equiv \{j_1, \cdots, j_k\}, \\ h^J(x) = (h^{j_1}(x), \cdots, h^{j_k}(x))^\top, \end{cases} \tag{3.2}$$

then $(g_x(\bar{x})^\top, h_x^J(\bar{x})^\top) \in \mathbb{R}^{n \times (m+k)}$ satisfies:

$$\text{rank}\,(g_x(\bar{x})^\top, h_x^J(\bar{x})^\top) = m + k. \tag{3.3}$$

Then there exist $\lambda \in \mathbb{R}^m$ and $\mu \in \mathbb{R}^\ell$ such that

$$\mu \geqslant 0, \tag{3.4}$$

$$\mu^\top h(\bar{x}) = 0, \tag{3.5}$$

$$f_x(\bar{x}) + \lambda^\top g_x(\bar{x}) + \mu^\top h_x(\bar{x}) = 0, \tag{3.6}$$

(ii) *Suppose $f(\cdot)$, $g(\cdot)$, and $h(\cdot)$ are C^2. Let $\bar{x}$ be a local optimal solution to Problem (3.1), at which the constraints are regular. Let*

$$L(x, \lambda, \mu) = f(x) + \lambda^\top g(x) + \mu^\top h(x), \qquad x \in \mathbb{R}^n. \tag{3.7}$$

Then

$$a^\top L_{xx}(\bar{x}, \lambda, \mu)a \geqslant 0, \qquad \forall a \in \mathcal{N}\left(\begin{pmatrix} g_x(\bar{x}) \\ h_x(\bar{x}) \end{pmatrix}\right). \tag{3.8}$$

(iii) *Suppose $f(\cdot)$, $g(\cdot)$, and $h(\cdot)$ are C^2 and for $\bar{x} \in \mathbb{R}^n$, there exist $\lambda \in \mathbb{R}^m$, $\mu \in \mathbb{R}^\ell$ with $g(\bar{x}) = 0$ and $h(\bar{x}) \leqslant 0$, $\mu \geqslant 0$ such that (3.5)–(3.6) hold. Moreover,*

$$a^\top L_{xx}(\bar{x}, \lambda, \mu)a > 0, \qquad \forall a \in \mathbb{M}, \ a \neq 0, \tag{3.9}$$

where

$$\mathbb{M} = \{a \in \mathbb{R}^n \mid g_x(\bar{x})a = 0, \quad h_x^j(\bar{x})a = 0, \ j \in \widehat{J}(\bar{x}, \mu)\}, \tag{3.10}$$

with

$$\widehat{J}(\bar{x}, \mu) \triangleq \{j \mid h^j(\bar{x}) = 0, \ \mu_j > 0, \ 1 \leqslant j \leqslant \ell\} \subseteq J(\bar{x}), \tag{3.11}$$

then $\bar{x}$ is a local minimum of Problem (3.1).

Part (i) of the above theorem is usually referred to as the Karush–Kuhn–Tucker (KKT, for short) Theorem which is due to W. Karush (1939), and H. W. Kuhn and A. W. Tucker (1951). This is also called the first-order necessary condition for the local minimum of Problem (3.1). Parts (ii) and (iii), which are due to McCormick (1967), are respectively called the second-order necessary and sufficient conditions for the local minimum of Problem (3.1).

Proof. (i) Let $\bar{x}$ be a local minimum of (3.1). Then $\bar{x}$ is also a local minimum of the following problem:

$$\begin{cases} \min & f(x), \\ \text{subject to} & g(x) = 0, \quad h^J(x) = 0. \end{cases} \tag{3.12}$$

This is an NLP problem with equality constraints. Without loss of generality, let us assume that

$$J(\bar{x}) = \{1, 2, \cdots, k\}, \quad h^J(x) = (h^1(x), \cdots, h^k(x))^\top.$$

Clearly, all the conditions assumed in Theorem 2.1 hold for Problem (3.12). Hence, applying Theorem 2.1, we can find $\lambda \in \mathbb{R}^m$ and $\mu^J \equiv (\mu_1, \cdots, \mu_k)^\top \in \mathbb{R}^k$ such that

$$f_x(\bar{x}) + \lambda^\top g_x(\bar{x}) + (\mu^J)^\top h_x^J(\bar{x}) = 0.$$

Now, we define

$$\mu = (\mu_1, \cdots, \mu_k, 0, \cdots, 0)^\top \in \mathbb{R}^\ell.$$

Then the (3.5) and (3.6) hold. We now need to show (3.4). To this end, we consider a map $\Phi : \mathbb{R}^n \times \mathbb{R}^k \to \mathbb{R}^{m+k}$ defined by the following:

$$\Phi(x, \eta) = \begin{pmatrix} g(x) \\ h^J(x) + \eta \end{pmatrix}, \qquad (x, \eta) \in \mathbb{R}^n \times \mathbb{R}^k.$$

Then

$$\Phi(\bar{x}, 0) = \begin{pmatrix} g(\bar{x}) \\ h^J(\bar{x}) \end{pmatrix} = 0,$$

and

$$\Phi_x(\bar{x}, 0) = \begin{pmatrix} g_x(\bar{x}) \\ h_x^J(\bar{x}) \end{pmatrix} \in \mathbb{R}^{(m+k) \times n},$$

which is of rank $(m + k)$. Hence, noting $n \geqslant m + k$, we have

$$\text{rank} \left(\Phi_x(\bar{x}, 0) \right) = m + k.$$

Without loss of generality, let $x = (\xi, \zeta) \in \mathbb{R}^{m+k} \times \mathbb{R}^{n-(m+k)}$, and $\bar{x} = (\bar{\xi}, \bar{\zeta})$ such that

$$\det \left(\Phi_\xi(\bar{x}, 0) \right) \equiv \det \left(\Phi_\xi(\bar{\xi}, \bar{\zeta}, 0) \right) \neq 0.$$

Then by Implicit Function Theorem, there exists a C^1 map $\varphi : \mathbb{R}^{n-(m+k)} \times \mathbb{R}^k \to \mathbb{R}^{m+k}$ such that

$$\Phi(\varphi(\zeta, \eta), \zeta, \eta) \equiv \begin{pmatrix} g(\varphi(\zeta, \eta), \zeta) \\ h^J(\varphi(\zeta, \eta), \zeta) + \eta \end{pmatrix} = 0, \qquad \forall (\zeta, \eta) \in U, \qquad (3.13)$$

with U being some neighborhood of $(\bar{\zeta}, 0)$, and

$$(\varphi(\bar{\zeta}, 0), \bar{\zeta}) = (\bar{\xi}, \bar{\zeta}) \equiv \bar{x}.$$

Differentiating equation (3.13) with respect to η, and then evaluating at $(\varphi(\bar{\zeta}, 0), \bar{\zeta}) = \bar{x}$, we obtain

$$\begin{cases} g_\xi(\bar{x}) \varphi_\eta(\bar{\zeta}, 0) = 0, \\ h_\xi^J(\bar{x}) \varphi_\eta(\bar{\zeta}, 0) + I = 0. \end{cases} \qquad (3.14)$$

Let

$$\widehat{h}(x) = (h^{k+1}(x), \cdots, h^\ell(x))^\top, \qquad x \in \mathbb{R}^n.$$

Then

$$\widehat{h}(\varphi(\bar{\zeta}, 0), \bar{\zeta}) = \widehat{h}(\bar{x}) < 0.$$

Thus, by shrinking U if necessary, we may assume that

$$\widehat{h}(\varphi(\zeta, \eta), \zeta) = \widehat{h}(x) \leqslant 0, \qquad \forall (\zeta, \eta) \in U.$$

The above, together with (3.13), implies that $(\varphi(\zeta, \eta), \eta)$ are feasible for all $(\zeta, \eta) \in U$ with $\eta \geqslant 0$. Consequently, by the optimality of $\bar{x}$,

$$f(\varphi(\bar{\zeta}, \eta), \bar{\zeta}) \geqslant f(\bar{x}) \equiv f(\bar{\xi}, \bar{\zeta}) = f(\varphi(\bar{\zeta}, 0), \bar{\zeta}),$$

for all small enough $\eta \geqslant 0$. Then for any small $\eta \geqslant 0$, (making use of (3.14) and the fact that $\mu_i = 0$ for $k + 1 \leqslant i \leqslant \ell$)

$$0 \leqslant f_\xi(\bar{x})\varphi_\eta(\bar{\zeta}, 0)\eta = -[\lambda^\top g_\xi(\bar{x}) + \mu^\top h_\xi(\bar{x})]\varphi_\eta(\bar{\zeta}, 0)\eta$$
$$= -(\mu^J)^\top h_\xi^J(\bar{x})\varphi_\eta(\bar{\zeta}, 0)\eta = (\mu^J)^\top \eta, \qquad \forall \eta \geqslant 0.$$

This implies that

$$\mu^J \geqslant 0,$$

proving (3.4).

(ii) By Theorem 2.1, we know that if one defines

$$\widetilde{L}(x, \lambda, \mu^J) = f(x) + \lambda^\top g(x) + (\mu^J)^\top h^J(x),$$

then

$$a^\top[\widetilde{L}_{xx}(\bar{x}, \lambda, \mu^J)]a \geqslant 0, \qquad \forall a \in \mathcal{N}\left(\begin{pmatrix} g_x(\bar{x}) \\ h_x^J(\bar{x}) \end{pmatrix}\right). \qquad (3.15)$$

Since $\mu_{k+1} = \cdots = \mu_\ell = 0$, we see that

$$\widetilde{L}(x, \lambda, \mu^J) = f(x) + \lambda^\top g(x) + \mu^\top h(x) = L(x, \lambda, \mu).$$

Also,

$$\mathcal{N}\left(\begin{pmatrix} g_x(\bar{x}) \\ h_x(\bar{x}) \end{pmatrix}\right) \subseteq \mathcal{N}\left(\begin{pmatrix} g_x(\bar{x}) \\ h_x^J(\bar{x}) \end{pmatrix}\right).$$

Hence, (3.8) follows from (3.15).

(iii) Suppose $\bar{x}$ is not a local minimum of Problem (3.1). Then we can find a sequence s_k, $\|s_k\| = 1$, and $\delta_k > 0$, $\delta_k \to 0$ such that $\bar{x} + \delta_k s_k$ is feasible for each $k \geqslant 1$ and

$$f(\bar{x} + \delta_k s_k) < f(\bar{x}), \qquad k \geqslant 1.$$

We may assume that $s_k \to \bar{s}$, with $\|\bar{s}\| = 1$. Clearly, the above implies

$$0 > f(\bar{x} + \delta_k s_k) - f(\bar{x}) = \delta_k f_x(\bar{x})s_k + \frac{\delta_k^2}{2}s_k^\top f_{xx}(\bar{x})s_k + o(\delta_k^2), \qquad (3.16)$$

which leads to

$$f_x(\bar{x})\bar{s} \leqslant 0. \qquad (3.17)$$

On the other hand, since $\bar{x} + \delta_k s_k$ is feasible,

$$0 = g_i(\bar{x} + \delta_k s_k) = g^i(\bar{x}) + \delta_k g_x^i(\bar{x})s_k + \frac{\delta_k^2}{2}s_k^\top g_{xx}^i(\bar{x})s_k + o(\delta_k^2). \qquad (3.18)$$

Moreover, for each $j \in J(\bar{x})$ (see (3.2)), we have (note $h^j(\bar{x}) = 0$)

$$0 \geqslant h^j(\bar{x} + \delta_k s_k) - h^j(\bar{x}) = \delta_k h_x^j(\bar{x})^\top s_k + \frac{\delta_k^2}{2} s_k^\top h_{xx}^j(\bar{x}) s_k + o(\delta_k^2). \quad (3.19)$$

Therefore, we have

$$g_x^i(\bar{x})\bar{s} = 0, \qquad 1 \leqslant i \leqslant m, \quad (3.20)$$

and

$$h_x^j(\bar{x})\bar{s} \leqslant 0, \qquad j \in J(\bar{x}). \quad (3.21)$$

Now, we claim that the equality in (3.21) holds for all $j \in \widehat{J}(\bar{x}, \mu) \subseteq J(\bar{x})$. In fact, if for some $j \in \widehat{J}(\bar{x}, \mu)$, we have

$$h_x^j(\bar{x})^\top \bar{s} < 0,$$

then (note (3.17)–(3.21) and (3.4)–(3.6))

$$0 \geqslant f_x(\bar{x})\bar{s} = -\left[\lambda^\top g_x(\bar{x}) + \mu^\top h_x(\bar{x})\right]\bar{s} = -\mu^\top h_x(\bar{x})\bar{s} > 0,$$

a contradiction. Hence, our claim holds, which means that $\bar{s} \in \mathbb{M}$. Now, by (3.16), (3.18) and (3.19), we obtain

$$0 > \delta_k \left[f_x(\bar{x}) + \lambda^\top g_x(\bar{x}) + \mu^\top h_x(\bar{x}) \right] s_k$$

$$+ \frac{\delta_k^2}{2} s_k^\top \left[f_{xx}(\bar{x}) + \sum_{i=1}^m \lambda_i g_{xx}^i(\bar{x}) + \sum_{j=1}^\ell \mu^j h_{xx}^j(\bar{x}) \right] s_k + o(\delta_k^2)$$

$$= \frac{\delta_k^2}{2} s_k^\top \left[f_{xx}(\bar{x}) + \sum_{i=1}^m \lambda_i g_{xx}^i(\bar{x}) + \sum_{j=1}^\ell \mu^j h_{xx}^j(\bar{x}) + o(1) \right] s_k$$

$$= \frac{\delta_k^2}{2} s_k^\top \left[L_{xx}(\bar{x}, \lambda, \mu) + o(1) \right] s_k.$$

This implies that

$$\bar{s}^\top L_{xx}(\bar{x}, \lambda, \mu) \bar{s} \leqslant 0,$$

contradicting the positive definiteness of $L_{xx}(\bar{x}, \lambda, \mu)$ on $\mathbb{M}$, proving the sufficiency. $\qquad \square$

For a local optimal solution $\bar{x}$ to Problem (3.1), we have

$$g(\bar{x}) = 0, \quad h(\bar{x}) \leqslant 0,$$

which is called the *primal feasibility*. From the above theorem, we should also have (3.6) which is called the *stationarity*, (3.4) which is called the *dual feasibility*, and (3.5) which is called the *complementary slackness*. We refer to (3.4)–(3.6) as the *first order Karush–Kuhn–Tucker conditions* (KKT conditions, for short), and refer to (3.8) as the *second order necessary KKT condition*, and to (3.9) as the *second order sufficient KKT condition*. We call (λ, μ) the *KKT multiplier*.

Now, let us look at the following problem:

$$\begin{cases} \max \quad f(x), \\ \text{subject to} \quad g(x) = 0, \quad h(x) \leqslant 0. \end{cases} \tag{3.22}$$

For this problem, we have the following result which is very similar to Theorem 3.3.

Theorem 3.4. *Let $f(\cdot)$, $g(\cdot)$, and $h(\cdot)$ be C^1.*

(i) *Let $\bar{x}$ be a local optimal solution to Problem (3.22) and the constraints be regular at $\bar{x}$. Then there exist $\lambda \in \mathbb{R}^m$ and $\mu \in \mathbb{R}^\ell$ such that*

$$\mu \leqslant 0, \tag{3.23}$$

$$\mu^\top h(\bar{x}) = 0, \tag{3.24}$$

$$f_x(\bar{x}) + \lambda^\top g_x(\bar{x}) + \mu^\top h_x(\bar{x}) = 0, \tag{3.25}$$

and if, in addition, $f(\cdot)$, $g(\cdot)$, and $h(\cdot)$ are C^2 and we denote $L(x, \lambda, \mu)$ by (3.7), then

$$a^\top L_{xx}(\bar{x}, \lambda, \mu)a \leqslant 0, \qquad \forall a \in \mathcal{N}\left(\begin{pmatrix} g_x(\bar{x}) \\ h_x(\bar{x}) \end{pmatrix} \right). \tag{3.26}$$

(ii) *Suppose $f(\cdot)$, $g(\cdot)$, and $h(\cdot)$ are C^2, and for $\bar{x} \in \mathbb{R}^n$, there exist $\lambda \in \mathbb{R}^m$, $\mu \in \mathbb{R}^\ell$, with $\mu \leqslant 0$ such that (3.24)–(3.25) hold and*

$$a^\top L_{xx}(\bar{x}, \lambda, \mu)a < 0, \qquad \forall a \in \mathbb{M}, \ a \neq 0, \tag{3.27}$$

where $\mathbb{M}$ is defined by (3.10) with $\mu_j > 0$ replaced by $\mu_j < 0$ in the definition of $\widehat{J}(\bar{x}, \bar{\mu})$. Then $\bar{x}$ is a local maximum of Problem (3.22).

The proof is left to the readers.

Note that when $\ell = 0$, Theorems 3.1 and 3.3 are reduced to Theorems 2.1 and 2.2.

Similar to the Lagrange multiplier method, for problems with equality and inequality constraint, we may define *KKT function*

$$L(x, \lambda, \mu) = f(x) + \lambda^\top g(x) + \mu^\top h(x), \quad (x, \lambda, \mu) \in \mathbb{R}^n \times \mathbb{R}^m \times \mathbb{R}^\ell, \tag{3.28}$$

and look for critical points of this function without constraint. If $(\bar{x}, \lambda, \mu)$ is a critical point of $L(\cdot, \cdot, \cdot)$, then $\bar{x}$ will be a candidate of optimal solution to Problem (3.1) or (3.22). Using either the second order KKT sufficient condition or some other method, such as Weierstrass Theorem, one may be able to determine if $\bar{x}$ is an optimal solution. Such a method is referred to as the KKT method.

We now present an example.

Example 3.5. Let

$$f(x, y, z) = x^2 + y^2 + z^2,$$
$$g(x, y, z) = (x - 1)^2 + (y - 1)^2 + (z - 1)^2 - 1,$$
$$h(x, y, z) = x + y + z - 2.$$

Find maximum and minimum of $f(x, y, z)$ subject to $g(x, y, z) = 0$ and $h(x, y, z) \leqslant 0$. To this end, we form the following KKT function

$$L(x, y, z, \lambda, \mu) = x^2 + y^2 + z^2 + \lambda[(x - 1)^2 + (y - 1)^2 + (z - 1)^2 - 1]$$
$$+ \mu(x + y + z - 2).$$

Set

$$0 = L_x(x, y, z, \lambda, \mu) = 2x + 2\lambda(x - 1) + \mu = 2(\lambda + 1)(x - 1) + \mu + 2,$$
$$0 = L_y(x, y, z, \lambda, \mu) = 2y + 2\lambda(y - 1) + \mu = 2(\lambda + 1)(y - 1) + \mu + 2,$$
$$0 = L_z(x, y, z, \lambda, \mu) = 2z + 2\lambda(z - 1) + \mu = 2(\lambda + 1)(z - 1) + \mu + 2.$$

Thus, by the equality constraint, one has

$$4(\lambda + 1)^2 = 4(\lambda + 1)^2 \left[(x - 1)^2 + (y - 1)^2 + (z - 1)^2 \right] = 3(\mu + 2)^2. \quad (3.29)$$

Now, we look at the minimum and maximum separately.

(i) For minimum, if $\lambda = -1$, then (3.29) implies

$$\mu = -2 < 0,$$

a contradiction. Hence, $\lambda \neq -1$, leading to

$$\frac{\mu + 2}{\lambda + 1} = \pm \frac{2}{\sqrt{3}},$$

and

$$x = y = z = 1 - \frac{\mu + 2}{2(\lambda + 1)} = \frac{2\lambda - \mu}{2(\lambda + 1)}.$$

Hence, we require $\mu \geqslant 0$, and

$$0 = \mu(x + y + z - 2) = \mu\left[\frac{6\lambda - 3\mu}{2(\lambda + 1)} - 2\right] = \frac{\mu(2\lambda - 3\mu - 4)}{2(1 + \lambda)}.$$

If $\mu \neq 0$, then

$$2\lambda - 3\mu - 4 = 0,$$

which gives

$$\lambda = \frac{3\mu + 4}{2}.$$

Then we have

$$\mu + 2 = \pm\frac{2}{\sqrt{3}}(\lambda + 1) = \pm\frac{2}{\sqrt{3}}\left[\frac{3\mu + 4}{2} + 1\right] = \pm\frac{3\mu + 6}{\sqrt{3}}.$$

Hence,

$$\mu = \frac{-2\sqrt{3} \pm 6}{\sqrt{3} \mp 3} = -2 < 0,$$

which contradicts $\mu > 0$. This means that $\mu = 0$ has to be true. Now, for $\mu = 0$, we have

$$\frac{1}{\lambda + 1} = \pm\frac{1}{\sqrt{3}},$$

which gives

$$\lambda = -1 \pm \sqrt{3},$$

and

$$x = y = z = \frac{\lambda}{\lambda + 1} = \frac{-1 \pm \sqrt{3}}{\mp\sqrt{3}} = 1 \mp \frac{\sqrt{3}}{3}.$$

In this case, we need

$$x + y + z - 2 = 3\left(1 \mp \frac{\sqrt{3}}{3}\right) - 2 = 1 \mp \sqrt{3} < 0.$$

Hence, the only candidate of the minimizer is

$$x = y = z = 1 - \frac{\sqrt{3}}{3}.$$

Since in the current case,

$$F = \Big\{(x, y, z) \mid g(x, y, z) = 0, \ h(x, y, z) \leqslant 0\Big\}$$

is compact, global minimum exists. Hence, $(1 - \frac{\sqrt{3}}{3})(1, 1, 1)^\top \in \mathbb{R}^3$ is the minimum.

(ii) From the above, we have seen that $\mu = 0$ will give us the minimum. Hence, for maximum, we should have $\mu \neq 0$. According to the above, if $\lambda = -1$, $\mu = -2$, which is allowed. Also, if $\lambda \neq -1$ and $\mu \neq 0$, then $\mu = -2$ as well. Thus, regardless of λ, as long as $\mu \neq 0$, one must have $\mu = -2$. Then by (3.29), we must have $\lambda = -1$. Note that for such λ and μ, we have

$$\nabla L(x, y, z, \lambda, \mu) = 0, \qquad \forall (x, y, z) \in \mathbb{R}^3.$$

Hence, any points are critical points.

More carefully, from $\mu = -2$, we know that the inequality constraint has to be active at the maximum point(s). This means that the maximum points have to satisfy the following:

$$(x - 1)^2 + (y - 1)^2 + (z - 1)^2 = 1, \quad x + y + z = 2. \tag{3.30}$$

Let (x, y, z) be any point satisfying the above equations. Then

$$x^2 + y^2 + z^2 = 2x + 2y + 2z - 2 = 2.$$

This means that the objective function is a constant on the set specified by the equations in (3.30). Hence, all the points on that set are maximum points. Geometrically, the set determined by (3.30) is a circle.

Similar to the previous section, we have the following summary about the KKT method.

• If optimal solutions do not exist, or the constraints are not regular at the optimal solution, the KKT method might not work.

• If one can determine that the feasible set is compact, or the objective function is coercive, then the optimal solution can be determined by the first order KKT conditions (the second sufficient condition is not necessary).

Sometimes, we might have problems with inequality constraints only (the equality constraints are absent). For such kind of problems, we have similar results as Theorems 3.1 and 3.3. More precisely, we allow $m = 0$ in Theorems 3.1 and 3.3. For such cases, the terms involving $g(\cdot)$ are all absent. We leave the details to the readers.

To conclude this section, let us present an interesting result. To this end, we first introduce the following definition.

Definition 3.6. A point $(\bar{x}, \bar{\lambda}, \bar{\mu}) \in \mathbb{R}^n \times \mathbb{R}^m \times \mathbb{R}_+^\ell$ is called a *saddle point* of $L(\cdot, \cdot, \cdot)$ (over $\mathbb{R}^n \times \mathbb{R}^m \times \mathbb{R}_+^\ell$) if the following holds:

$$L(\bar{x}, \lambda, \mu) \leqslant L(\bar{x}, \bar{\lambda}, \bar{\mu}) \leqslant L(x, \bar{\lambda}, \bar{\mu}), \quad \forall (x, \lambda, \mu) \in \mathbb{R}^n \times \mathbb{R}^m \times \mathbb{R}_+^\ell.$$

We have the following result.

Theorem 3.7. *Let $(\bar{x}, \bar{\lambda}, \bar{\mu}) \in \mathbb{R}^n \times \mathbb{R}^m \times \mathbb{R}_+^\ell$ be a saddle point of $L(\cdot, \cdot, \cdot)$. Then $\bar{x}$ is an optimal solution to Problem* (3.1).

Proof. We first show that $\bar{x}$ is feasible. In fact, from

$$f(\bar{x}) + \lambda^\top g(\bar{x}) + \mu^\top h(\bar{x}) \leqslant f(\bar{x}) + \bar{\lambda}^\top g(\bar{x}) + \bar{\mu}^\top h(\bar{x}),$$

we have

$$\lambda^\top g(\bar{x}) + \mu^\top h(\bar{x}) \leqslant \bar{\lambda}^\top g(\bar{x}) + \bar{\mu}^\top h(\bar{x}), \qquad \forall (\lambda, \mu) \in \mathbb{R}^m \times \mathbb{R}_+^\ell.$$

Thus, it is necessary that

$$g(\bar{x}) = 0, \qquad h(\bar{x}) \leqslant 0,$$

proving the feasibility of $\bar{x}$. Also, the above implies that

$$\mu^\top h(\bar{x}) \leqslant \bar{\mu}^\top h(\bar{x}) \leqslant 0, \qquad \forall \mu \in \mathbb{R}_+^\ell,$$

which leads to

$$\bar{\mu}^\top h(\bar{x}) = 0.$$

Consequently, from

$$f(\bar{x}) + \bar{\lambda}^\top g(\bar{x}) + \bar{\mu}^\top h(\bar{x}) \leqslant f(x) + \bar{\lambda}^\top g(x) + \bar{\mu}^\top h(x), \qquad \forall x \in \mathbb{R}^n,$$

it follows that

$$f(\bar{x}) \leqslant f(x), \qquad \forall x \in \{x \in \mathbb{R}^n \mid g(x) = 0, \ h(x) \leqslant 0\},$$

proving the optimality of $\bar{x}$. □

The following result is closely related to the above.

Proposition 3.8. *A necessary and sufficient condition for a point $(\bar{x}, \bar{\lambda}, \bar{\mu}) \in \mathbb{R}^n \times \mathbb{R}^m \times \mathbb{R}_+^\ell$ to be a saddle point for $L(x, \lambda, \mu)$ is the following:*

$$\begin{cases} L(\bar{x}, \bar{\lambda}, \bar{\mu}) = \min_{x \in \mathbb{R}^n} L(x, \bar{\lambda}, \bar{\mu}), \\ g(\bar{x}) = 0, \quad h(\bar{x}) \leqslant 0, \\ \bar{\mu}^\top h(\bar{x}) = 0. \end{cases} \qquad (3.31)$$

Moreover, in this case,

$$f(\bar{x}) = L(\bar{x}, \bar{\lambda}, \bar{\mu}). \qquad (3.32)$$

Proof. Necessity. Let $(\bar{x}, \bar{\lambda}, \bar{\mu}) \in \mathbb{R}^n \times \mathbb{R}^m \times \mathbb{R}^\ell_+$ be a saddle point of $L(x, \lambda, \mu)$. Then the first equality in (3.31) follows from the definition of saddle point. The last two relations in (3.31) follows from the proof of Theorem 3.7. It is clear that under the last two relations in (3.31), one has
$$L(\bar{x}, \bar{\lambda}, \bar{\mu}) = f(\bar{x}) + \bar{\lambda}^\top g(\bar{x}) + \bar{\mu}^\top h(\bar{x}) = f(\bar{x}),$$
which gives (3.32).

Sufficiency. Suppose (3.31) holds, which implies (3.32). Then the first relation in (3.31) and (3.32) imply
$$f(\bar{x}) = L(\bar{x}, \bar{\lambda}, \bar{\mu}) \leqslant L(x, \bar{\lambda}, \bar{\mu}), \qquad \forall x \in \mathbb{R}^n,$$
and for all $(\lambda, \mu) \in \mathbb{R}^m \times \mathbb{R}^\ell_+$, by the feasibility of $\bar{x}$,
$$L(\bar{x}, \lambda, \mu) = f(\bar{x}) + \lambda^\top g(\bar{x}) + \mu^\top h(\bar{x}) \leqslant f(\bar{x}) = L(\bar{x}, \bar{\lambda}, \bar{\mu}),$$
which means that $(\bar{x}, \bar{\lambda}, \bar{\mu})$ is a saddle point of $L(x, \lambda, \mu)$. □

Exercises

1. Prove Theorem 3.3. (Hint: Using Theorem 3.1)

2. Let $f : \mathbb{R}^n \to \mathbb{R}$ and $h : \mathbb{R}^n \to \mathbb{R}^\ell$. State and prove a similar theorem for the following problem:
$$\begin{cases} \min & f(x), \\ \text{subject to} & h(x) \leqslant 0. \end{cases}$$

3. Let $f : \mathbb{R}^n \to \mathbb{R}$ and $h : \mathbb{R}^n \to \mathbb{R}^\ell$. State and prove a similar theorem for the following problem:
$$\begin{cases} \max & f(x), \\ \text{subject to} & h(x) \leqslant 0. \end{cases}$$

4. Let $f(x, y, z) = xyz$, $g(x, y, z) = x + y + z - 1$, and $h(x, y, z) = x^2 + y^2 + z^2 - 1$. Find minimum and maximum of $f(x, y, z)$ subject to $g(x, y, z) = 0$ and $h(x, y, z) \leqslant 0$. Try to extend to the case of n variables.

5. Solve the following problem:
$$\begin{cases} \max & \ln x + \ln y, \\ \text{subject to} & x^2 + y^2 = 1, \qquad x, y \geqslant 0. \end{cases}$$

6. Let $f : \mathbb{R}^n \to \mathbb{R}$ and $h : \mathbb{R}^n \to \mathbb{R}$ be given by $f(x) = \|x\|^2$, $h(x) = a^\top x - 1$, with $a \in \mathbb{R}^n$ and $\|a\| = 1$. Find the minimum of $f(x)$ subject to $h(x) \leqslant 0$.

7. Find solutions to the following problem:
$$\begin{cases} \min & x^2 + (y+1)^2, \\ \text{subject to} & x^2 + y^2 \leqslant 4, \quad y^2 \leqslant x - 1, \quad -2y \leqslant x - 1. \end{cases}$$

4   Problems without Regularity of the Constraints

We see that in the previous sections, the regularity of the constraints at the local optimal solution $\bar{x}$ plays an essential role. A natural question is what if we do not know such a condition holds *a priori*? Can we still have certain necessary and sufficient conditions for the (local) optimal solutions? In this section we would like to use a different approach to establish the first order necessary conditions for optimal solutions to NLP problems, without assuming the regularity condition. The main result is due to Fritz John (1948). Our proof is based on an interesting result due to Ekeland (1974).

Lemma 4.1. (Ekeland's Variational Principle) *Let $G \subseteq \mathbb{R}^n$ be closed and $\varphi : G \to \mathbb{R}$ be lower semi-continuous, bounded from below. Let $x_0 \in G$ and $\delta > 0$ be fixed. Then there exists an $\bar{x} \in G$, such that*

$$\varphi(\bar{x}) + \delta\|\bar{x} - x_0\| \leqslant \varphi(x_0), \tag{4.1}$$

$$\varphi(\bar{x}) < \varphi(x) + \delta\|x - \bar{x}\|, \qquad \forall x \neq \bar{x}, \ x \in G. \tag{4.2}$$

Proof. By considering $\varphi(\cdot) - \inf_{x \in G} \varphi(x)$ instead of $\varphi(\cdot)$, we may assume that $\varphi(\cdot)$ is non-negative valued. Next, we define

$$\Gamma(x) = \{y \in G \mid \varphi(y) + \delta\|x - y\| \leqslant \varphi(x)\}.$$

Then, for any $x \in G$, $\Gamma(x)$ is a non-empty closed set since $x \in \Gamma(x)$. Next, we claim that

$$y \in \Gamma(x) \quad \Rightarrow \quad \Gamma(y) \subseteq \Gamma(x). \tag{4.3}$$

In fact, $y \in \Gamma(x)$ implies

$$\varphi(y) + \delta\|x - y\| \leqslant \varphi(x),$$

and for any $z \in \Gamma(y)$, we have

$$\varphi(z) + \delta\|y - z\| \leqslant \varphi(y).$$

Thus, using the triangle inequality, one has

$$\varphi(z) + \delta\|x - z\| \leqslant \varphi(z) + \delta\|y - z\| + \delta\|x - y\|$$
$$\leqslant \varphi(y) + \delta\|x - y\| \leqslant \varphi(x).$$

Hence, (4.3) holds. Next, we define

$$\bar{\varphi}(x) = \inf_{y \in \Gamma(x)} \varphi(y), \qquad \forall x \in G.$$

Then, for any $y \in \Gamma(x)$, $\bar{\varphi}(x) \leqslant \varphi(y) \leqslant \varphi(x) - \delta \|x - y\|$. Hence,

$$\delta \|x - y\| \leqslant \varphi(x) - \bar{\varphi}(x).$$

Thus, the diameter $\operatorname{diam} \Gamma(x)$ of the set $\Gamma(x)$ satisfies

$$\operatorname{diam} \Gamma(x) \equiv \sup_{y, z \in \Gamma(x)} \|y - z\| \leqslant \frac{2}{\delta} [\varphi(x) - \bar{\varphi}(x)]. \tag{4.4}$$

This implies that each $\Gamma(x)$ is bounded and closed (and thus it is compact). Now, by the compactness of $\Gamma(x_0)$, we can find an $\bar{x} \in \Gamma(x_0)$ such that

$$\varphi(\bar{x}) = \inf_{x \in \Gamma(x_0)} \varphi(x) \equiv \bar{\varphi}(x_0).$$

Then by (4.4), we obtain (note $\Gamma(\bar{x}) \subseteq \Gamma(x_0)$)

$$0 \leqslant \operatorname{diam} \Gamma(\bar{x}) \leqslant \frac{2}{\delta} [\varphi(\bar{x}) - \bar{\varphi}(\bar{x})] = \frac{2}{\delta} \left[\inf_{x \in \Gamma(x_0)} \varphi(x) - \inf_{x \in \Gamma(\bar{x})} \varphi(x) \right] \leqslant 0.$$

Thus, $\Gamma(\bar{x}) = \{\bar{x}\}$. Consequently, any $x \neq \bar{x}$ is not in $\Gamma(\bar{x})$, which gives (4.2). $\qquad\square$

Corollary 4.2. *Let the assumption of Lemma 4.1 hold. Let $\varepsilon > 0$ and $x_0 \in G$ be such that*

$$\varphi(x_0) \leqslant \inf \varphi(G) + \varepsilon.$$

Then, there exists an $x_\varepsilon \in G$ such that

$$\varphi(x_\varepsilon) \leqslant \varphi(x_0), \qquad \|x_\varepsilon - x_0\| \leqslant \sqrt{\varepsilon}, \tag{4.5}$$

and

$$\varphi(x_\varepsilon) - \sqrt{\varepsilon} \|x - x_\varepsilon\| \leqslant \varphi(x), \qquad \forall x \in G. \tag{4.6}$$

Proof. We take $\delta = \sqrt{\varepsilon}$ in Lemma 4.1. Then there exists an $x_\varepsilon \in G$ such that

$$\varphi(x_\varepsilon) + \sqrt{\varepsilon} \|x_\varepsilon - x_0\| \leqslant \varphi(x_0) \leqslant \inf \varphi(G) + \varepsilon \leqslant \varphi(x_\varepsilon) + \varepsilon,$$

and

$$\varphi(x_\varepsilon) < \varphi(x) + \sqrt{\varepsilon} \|x - x_\varepsilon\|, \qquad \forall x \neq x_\varepsilon.$$

Then, (4.5) and (4.6) follow immediately. $\qquad\square$

Note that G is not necessarily compact. Thus, $\varphi(\cdot)$ might not achieve its infimum over G. Corollary 4.2 says that if x_0 is an approximate minimum of $\varphi(\cdot)$ over G, then by making a small perturbation to the function $\varphi(\cdot)$, a minimum will be achieved at a point x_ε which is very close to x_0.

Now, we are ready to prove the following first order necessary conditions of Fritz John for optimal solution of Problem (3.1).

Theorem 4.3. (Fritz John) *Let $f(\cdot)$, $g(\cdot)$ and $h(\cdot)$ be C^1. Let $\bar{x} \in F$ be an optimal solution of Problem (3.1). Then there exists a non-zero vector $(\lambda^0, \lambda, \mu) \in [0,1] \times \mathbb{R}^m \times \mathbb{R}^\ell_+$ such that*

$$\lambda^0 f_x(\bar{x}) + \lambda^\top g_x(\bar{x}) + \mu^\top h_x(\bar{x}) = 0. \tag{4.7}$$

Let

$$J(\bar{x}) = \{j \mid 1 \leqslant j \leqslant \ell,\ h^j(\bar{x}) = 0\}. \tag{4.8}$$

Then

$$\mu_j = 0, \qquad \forall j \notin J(\bar{x}). \tag{4.9}$$

Proof. Suppose $\bar{x}$ is an optimal solution of (3.1). For any $\varepsilon > 0$, define

$$f^\varepsilon(x) = \Big\{ \big[\big(f(x) - f(\bar{x}) + \varepsilon\big)^+ \big]^2 + |g(x)|^2 + \sum_{j=1}^\ell [h^j(x)^+]^2 \Big\}^{\frac{1}{2}}, \quad x \in \mathbb{R}^n.$$

Clearly,

$$f^\varepsilon(x) > 0, \qquad \forall x \in \mathbb{R}^n,$$

and

$$f^\varepsilon(\bar{x}) = \varepsilon \leqslant \inf_{x \in \mathbb{R}^n} f^\varepsilon(x) + \varepsilon.$$

Thus, by Corollary 4.2, there exists an x^ε such that

$$\|x^\varepsilon - \bar{x}\| < \sqrt{\varepsilon},$$

and

$$f^\varepsilon(x^\varepsilon) - \sqrt{\varepsilon}\|x^\varepsilon - x\| \leqslant f^\varepsilon(x), \qquad \forall x \in \mathbb{R}^n.$$

Now, for any $y \in \mathbb{R}^n$ with $\|y\| = 1$, and $\delta > 0$, take $x = x^\varepsilon + \delta y$ in the above, we obtain

$$-\sqrt{\varepsilon} \leqslant \frac{f^\varepsilon(x^\varepsilon + \delta y) - f^\varepsilon(x^\varepsilon)}{\delta} = \frac{f^\varepsilon(x^\varepsilon + \delta y)^2 - f^\varepsilon(x^\varepsilon)^2}{[f^\varepsilon(x^\varepsilon + \delta y) + f^\varepsilon(x^\varepsilon)]\delta}$$

$$= \frac{1}{[f^\varepsilon(x^\varepsilon + \delta y) + f^\varepsilon(x^\varepsilon)]\delta} \Big\{ \big[\big(f(x^\varepsilon + \delta y) - f(\bar{x}) + \varepsilon\big)^+ \big]^2$$

$$- \big[\big(f(x^\varepsilon) - f(\bar{x}) + \varepsilon\big)^+ \big]^2 + |g(x^\varepsilon + \delta y)|^2 - |g(x^\varepsilon)|^2$$

$$+ \sum_{j=1}^\ell \Big([h^j(x^\varepsilon + \delta y)^+]^2 - [h^j(x^\varepsilon)^+]^2 \Big) \Big\}$$

$$\to \frac{1}{f^\varepsilon(x^\varepsilon)} \Big\{ \big(f(x^\varepsilon) - f(\bar{x}) + \varepsilon \big)^+ f_x(x^\varepsilon)y$$

$$+ \sum_{i=1}^m g^i(x^\varepsilon) g_x^i(x^\varepsilon)y + \sum_{j=1}^\ell h^j(x^\varepsilon)^+ h_x^j(x^\varepsilon)y \Big\}, \quad \text{as } \delta \to 0.$$

If we denote

$$
\begin{cases}
\lambda^{0,\varepsilon} = \dfrac{\big(f(x^\varepsilon) - f(\bar{x}) + \varepsilon\big)^+}{f^\varepsilon(x^\varepsilon)} \geqslant 0\,, \\[2ex]
\lambda_i^\varepsilon = \dfrac{g^i(x^\varepsilon)}{f^\varepsilon(x^\varepsilon)}\,, \quad 1 \leqslant i \leqslant m, \\[2ex]
\mu_j^\varepsilon = \dfrac{h^j(x^\varepsilon)^+}{f^\varepsilon(x^\varepsilon)} \geqslant 0\,, \quad 1 \leqslant j \leqslant \ell,
\end{cases}
$$

then

$$
(\lambda^{0,\varepsilon})^2 + \sum_{i=1}^m (\lambda_i^\varepsilon)^2 + \sum_{j=1}^\ell (\mu_j^\varepsilon)^2 = 1, \tag{4.10}
$$

and

$$
-\sqrt{\varepsilon} \leqslant \Big[\lambda^{0,\varepsilon} f_x(x^\varepsilon) + \sum_{i=1}^m \lambda_i^\varepsilon g_x^i(x^\varepsilon) + \sum_{j=1}^\ell \mu_j^\varepsilon h_x^j(x^\varepsilon)\Big] y. \tag{4.11}
$$

By (4.10), we may assume that

$$
\begin{cases}
\lambda^{0,\varepsilon} \to \lambda^0, \\
\lambda^\varepsilon \equiv (\lambda_1^\varepsilon, \cdots, \lambda_m^\varepsilon) \to (\lambda_1, \cdots, \lambda_m) \equiv \lambda, \qquad \text{as } \varepsilon \to 0, \\
\mu^\varepsilon \equiv (\mu_1^\varepsilon, \cdots, \mu_\ell^\varepsilon) \to (\mu_1, \cdots, \mu_\ell) \equiv \mu,
\end{cases}
$$

with

$$
(\lambda^0)^2 + \|\lambda\|^2 + \|\mu\|^2 = 1, \quad \lambda^0 \geqslant 0, \ \mu \geqslant 0.
$$

Thus, $(\lambda^0, \lambda, \mu) \neq 0$. Next, sending $\varepsilon \to 0$ in (4.11), we obtain

$$
0 \leqslant \Big[\lambda^0 f_x(\bar{x}) + \lambda^\top g(\bar{x}) + \mu^\top h(\bar{x})\Big] y, \quad \forall y \in \mathbb{R}^n.
$$

Hence, (4.7) follows.

Finally, for any $j \notin J(\bar{x})$, by the convergence of x^ε to $\bar{x}$, we know that

$$
\mu_j^\varepsilon \equiv \frac{h^j(x^\varepsilon)^+}{f^\varepsilon(x^\varepsilon)} = 0, \qquad \text{for } \varepsilon > 0 \text{ small enough.}
$$

This implies (4.9). □

We define the *Fritz John's function* (or *generalized Lagrange function*) $\Lambda(x, \lambda^0, \lambda, \mu)$ as follows:

$$
\begin{aligned}
\Lambda(x, \lambda^0, \lambda, \mu) &= \lambda^0 f(x) + \lambda^\top g(x) + \mu^\top h(x), \\
&\forall (x, \lambda^0, \lambda, \mu) \in G \times \mathbb{R} \times \mathbb{R}^m \times \mathbb{R}_+^\ell.
\end{aligned} \tag{4.12}
$$

Then condition (4.7) becomes

$$\Lambda_x(\bar{x}, \lambda^0, \lambda, \mu) = 0. \tag{4.13}$$

Also, we see that condition (4.9) is equivalent to

$$\mu^{\mathsf{T}} h(\bar{x}) = 0.$$

In the above theorem, the triple $(\lambda^0, \lambda, \mu)$ is called a *Fritz John's multiplier*.

The following example show that sometimes, $\lambda^0 = 0$ is possible.

Example 4.4. Let $G = \mathbb{R}^2$, $f(x, y) = -x$, $g(x, y) = (x - 1)^3 + y$, and

$$h(x, y) = \begin{pmatrix} -x \\ -y \end{pmatrix}.$$

From the equality constraint, we have $y = -(x - 1)^3$. Then the inequality constraints tell us

$$x \geqslant 0, \qquad y = (1 - x)^3 \geqslant 0,$$

which leads to $0 \leqslant x \leqslant 1$. Consequently, the optimal solution is $(1, 0)$. We have

$$h(1, 0) = \begin{pmatrix} -1 \\ 0 \end{pmatrix}.$$

Thus, $J = \{2\}$. We also have

$$\begin{cases} \nabla f(1, 0) = (-1, 0), & \nabla g(1, 0) = (0, 1), \\ \nabla h(1, 0) = \begin{pmatrix} -1 & 0 \\ 0 & -1 \end{pmatrix}. \end{cases}$$

If $(\lambda^0, \lambda, \mu)$ is the corresponding Fritz John's multiplier, then

$$0 = \mu^{\mathsf{T}} h(1, 0) = (\mu_1, \mu_2) \begin{pmatrix} -x \\ -y \end{pmatrix} \bigg|_{x=1, y=0} = -\mu_1,$$

and

$$\begin{aligned} 0 &= \lambda^0 \nabla f(1, 0) + \lambda \nabla g(1, 0) + \mu^{\mathsf{T}} \nabla h(1, 0) \\ &= \lambda^0(-1, 0) + \lambda(0, 1) + \mu_1(-1, 0) + \mu_2(0, -1) \\ &= \lambda^0(-1, 0) + \lambda(0, 1) + \mu_2(0, -1) = (-\lambda^0, \lambda - \mu_2). \end{aligned}$$

Hence, we have $\lambda^0 = 0$, and $\lambda = \mu_2$. Note that in the current case, $\nabla g(1, 0)$ and $\nabla h^2(0, 1)$ are linearly dependent. Therefore, the constraints are not regular at $(1, 0)$.

For this minimization problem, let us start from the Fritz John's function

$$\Lambda(x, y, \lambda^0, \lambda, \mu_1, \mu_2) = -\lambda^0 x + \lambda[(x-1)^3 + y] - \mu_1 x - \mu_2 y.$$

Set

$$\begin{cases} 0 = \Lambda_x(x, y, \lambda^0, \lambda, \mu_1, \mu_2) = -\lambda^0 + 3\lambda(x-1)^2 - \mu_1, \\ 0 = \Lambda_y(x, y, \lambda^0, \lambda, \mu_1, \mu_2) = \lambda - \mu_2, \end{cases} \tag{4.14}$$

and (noting the equality constraint)

$$0 = \mu_1 h^1(x, y) = -\mu_1 x, \qquad 0 = \mu_2 h^2(x, y) = -\mu_2 y = \mu_2(x-1)^3.$$

From (4.14), we see that $\lambda = \mu_2 > 0$ (otherwise, we will end up with $\lambda^0 + \mu_1 = 0$ and then $(\lambda^0, \lambda, \mu_1, \mu_2) = 0$). Therefore, $x = 1$ and $\mu_1 = 0$, which further leads to $\lambda^0 = 0$ and $y = 0$. Hence, we obtain the unique candidate $(\bar{x}, \bar{y}) = (1, 0)$ for the minimum, and the corresponding Fritz John's multiplier $(\lambda^0, \lambda, \mu_1, \mu_2) = (0, \mu_2, 0, \mu_2)$. Since in the current case, we have the compactness of the feasible set F, minimum exists. Thus, $(\bar{x}, \bar{y}) = (1, 0)$ is the minimum.

In general, if $\lambda^0 = 0$, condition (4.7) becomes

$$\lambda^\top g_x(\bar{x}) + \mu^\top h_x(\bar{x}) = 0, \tag{4.15}$$

which does not involve the objective function $f(\cdot)$. In this case, it seems to be difficult to directly determine if the obtained candidates are optimal solutions to the problem. Hence, we hope to find conditions under which $\lambda^0 > 0$. When this is the case, we say that the Fritz John's multiplier $(\lambda^0, \lambda, \mu)$ is *qualified*. Note that in the case $\lambda^0 > 0$, by scaling, we can assume that $\lambda^0 = 1$. Clearly,

$$\Lambda(x, 1, \lambda, \mu) = L(x, \lambda, \mu), \qquad \forall (x, \lambda, \mu) \in G \times \mathbb{R}^m \times \mathbb{R}^\ell_+.$$

Hence, in the case $\lambda^0 = 1$, (4.7) becomes KKT condition

$$L_x(\bar{x}, \lambda, \mu) = f_x(\bar{x}) + \lambda^\top g_x(\bar{x}) + \mu^\top h_x(\bar{x}) = 0.$$

Clearly, if $\lambda_0 = 0$, $\{g_x^i(\bar{x}), h_x^j(\bar{x}) : 1 \leqslant i \leqslant m, \ j \in J(\bar{x})\}$ are linearly independent, i.e. the constraints are regular at $\bar{x}$, by (4.9), we see that (4.15) is impossible. Hence, $\lambda^0 > 0$ must be true. This recovers Theorems 3.3 and 3.4. We now would like to look at a little more general situations.

Definition 4.5. The functions $g(\cdot)$ and $h(\cdot)$ are said to satisfy the *MF qualification condition* at a feasible point $\bar{x}$ if $g_x^1(\bar{x}), \cdots, g_x^m(\bar{x})$ are linearly independent and there exists a $z \in \mathbb{R}^n$ such that

$$g_x(\bar{x})z = 0, \qquad h_x^J(\bar{x})z < 0,$$

where $h^J(\cdot)$ is defined in (3.2).

The MF qualification condition was introduced by Mangasarian and Fromovitz in 1967. We have the following result.

Theorem 4.6. *Let $g(\cdot)$ and $h(\cdot)$ satisfy the MF qualification condition at the optimal solution $\bar{x}$ of Problem (3.1). Then the Fritz John's multiplier is qualified.*

Proof. Suppose $\lambda^0 = 0$. Then we have (4.15). Since $\{g_x^1(\bar{x}), \cdots, g_x^m(\bar{x})\}$ are linearly independent, we must have $\mu \neq 0$. Further, since $\mu^j = 0$ for any $j \notin J(\bar{x})$, we may rewrite (4.15) as follows

$$\lambda^\top g_x(\bar{x}) + \mu_J^\top h_x^J(\bar{x}) = 0.$$

Now, multiplier the z in our condition, we have

$$0 = \lambda^\top g_x(\bar{x})z + \mu_J^\top h_x^J(\bar{x})z = \mu_J^\top h_x^J(\bar{x})z < 0,$$

since at least one component in μ_J is strictly positive, a contradiction. □

Let us present an example.

Example 4.7. Let $g(x,y) = x^2 + y^2 - 1$, $h(x,y) = x + y - b$, with $b > 0$. Then

$$\nabla g(x,y) = (2x, 2y), \qquad \nabla h(x,y) = (1,1).$$

Suppose $(\bar{x}, \bar{y})^\top$ is feasible such that

$$0 = h(\bar{x}, \bar{y}) = \bar{x} + \bar{y} - b.$$

Then, substituting $\bar{y} = b - \bar{x}$ into the equality constraint

$$0 = g(\bar{x}, \bar{y}) = \bar{x}^2 + \bar{y}^2 - 1 = \bar{x}^2 + (b - \bar{x})^2 - 1 = 2\bar{x}^2 - 2b\bar{x} + b^2 - 1.$$

Hence,

$$\bar{x} = \frac{2b \pm \sqrt{4b^2 - 8(b^2 - 1)}}{4} = \frac{b \pm \sqrt{2 - b^2}}{2}.$$

Therefore, we need $b \leqslant \sqrt{2}$. In this case,

$$\bar{y} = b - \bar{x} = \frac{b \mp \sqrt{2 - b^2}}{2}.$$

In order the constraints to satisfy the constraint qualification at the feasible point $(\bar{x}, \bar{y})^\top$, it suffices to have the existence of some (ξ, η) such that

$$\begin{cases} 0 = (\bar{x}, \bar{y}) \begin{pmatrix} \xi \\ \eta \end{pmatrix} = \bar{x}\xi + \bar{y}\eta, \\ (1, 1) \begin{pmatrix} \xi \\ \eta \end{pmatrix} = \xi + \eta < 0. \end{cases}$$

From the first equation, we have

$$\eta = -\frac{\bar{x}}{\bar{y}}\xi.$$

Plugging into the second relation, we have

$$0 > \xi + \eta = \frac{\bar{y} - \bar{x}}{\bar{y}}\xi = 2\frac{\mp\sqrt{2 - b^2}}{b \mp \sqrt{2 - b^2}}\xi.$$

Hence, we need only exclude the case $b = \sqrt{2}$. In another word, for $b \in (0, \sqrt{2})$, the constraints satisfy the MF qualification condition.

The following result is concerned with the MF qualification condition and the regularity of the constraints $g(x) = 0$, $h(x) \leqslant 0$.

Proposition 4.8. *Suppose $G \subseteq \mathbb{R}^n$ is a domain, $g : G \to \mathbb{R}^m$ and $h : G \to \mathbb{R}$ are C^1.*

(i) *If the constraints are regular at $\bar{x}$, then the MF qualification condition holds at $\bar{x}$.*

(ii) *If $\ell = 1$ and the MF qualification condition holds at $\bar{x}$, then the constraints are regular at $\bar{x}$.*

Proof. (i) If the constraints are regular at $\bar{x}$. Then with the previously defined J and $h^J(\cdot)$, one has that

$$\text{rank} \begin{pmatrix} g_x(\bar{x}) \\ h_x^J(\bar{x}) \end{pmatrix}_{(m+k) \times n} = m + k \leqslant n.$$

Hence, there exists a matrix $B \in \mathbb{R}^{(n-m-k) \times n}$ such that

$$\begin{pmatrix} B \\ g_x(\bar{x}) \\ h_x^J(\bar{x}) \end{pmatrix} z = \begin{pmatrix} 0 \\ 0 \\ -1 \end{pmatrix},$$

admits a unique solution z. Here, B is chosen so that the coefficient matrix of the above linear equation is invertible and $\mathbf{1} = (1, 1, \cdots, 1)^\top \in \mathbb{R}^{n-m-k}$. With this solution z, we have

$$g_x(\bar{x})z = 0, \quad h_x^J(\bar{x})z = -\mathbf{1} < 0.$$

This means that the MF qualification condition holds.

(ii) Now, let $\ell = 1$. We prove our conclusion by contradiction. Suppose the constraints are not regular at a feasible point $\bar{x}$. Then vectors

$g_x^1(\bar{x}), \cdots, g_x^m(\bar{x}), h_x(\bar{x})$ are linearly dependent. Thus, there exist $\lambda \in \mathbb{R}^m$ and $\mu \in \mathbb{R}$ such that

$$\lambda^\top g_x(\bar{x}) + \mu h_x(\bar{x}) = 0. \tag{4.16}$$

Since the MF qualification condition is assumed, there exists a $z \in \mathbb{R}^n$ such that

$$g_x(\bar{x})z = 0, \qquad h_x(\bar{x})z < 0, \tag{4.17}$$

and $g_x^1(\bar{x}), \cdots, g_x^m(\bar{x})$ are linearly independent. Hence, $\mu \neq 0$. Now, combining (4.16)–(4.17), one obtains

$$0 = \lambda^\top g_x(\bar{x})z + \mu h_x(\bar{x})z = \mu h_x(\bar{x})z \neq 0,$$

which is a contradiction. Thus, the constraints must be regular. $\qquad\square$

We point out that if $\ell > 1$, the MF qualification condition does not necessarily imply the regularity of the constraints. Therefore, the former is strictly weaker than the latter. Here is such an example.

Example 4.9. Consider

$$\begin{cases} \min \quad f(x) = (x_1 - 1)^2 + (x_2 - 1)^2 + (x_3 - 2)^2 + (x_4 - 2)^2, \\ \text{subject to} \quad g(x) = \dfrac{x_1^2}{2} + \dfrac{x_2^2}{2} + \dfrac{3x_3^2}{4} + \dfrac{3x_4^2}{4} - 7 = 0, \\ h^1(x) = x_1^2 + x_2^2 + x_3^2 + x_4^2 - 10 \leqslant 0, \\ h^2(x) = x_1 + x_2 + x_3 + x_4 - 6 \leqslant 0. \end{cases}$$

Then $n = 4$, $m = 1$, and $\ell = 2$. The optimal solution is given by

$$\bar{x}^\top \equiv (\bar{x}_1, \bar{x}_2, \bar{x}_3, \bar{x}_4) = (1, 1, 2, 2),$$

with $f(\bar{x}) = 0$, and

$$h^1(\bar{x}) = 0, \qquad h^2(\bar{x}) = 0.$$

Thus, $J = \{1, 2\}$. Now,

$$g_x(\bar{x}) = (1, 1, 3, 3),$$
$$h_x^1(\bar{x}) = (2, 2, 4, 4),$$
$$h_x^2(\bar{x}) = (1, 1, 1, 1).$$

Clearly, the above three vectors are linear dependent. Hence, the constraints are not regular at $\bar{x}$. However, by taking

$$z^\top \equiv (z_1, z_2, z_3, z_4) = (-3, -3, 1, 1),$$

we see that

$$g_x(\bar{x})z = 0, \quad h_x^1(\bar{x})z = -4 < 0, \quad h_x^2(\bar{x})z = -4 < 0.$$

This means that the MF qualification condition is satisfied.

For the problem with the equality constraints being absent, we have another interesting notion.

Definition 4.10. Vectors $\xi_1, \cdots, \xi_k \in \mathbb{R}^n$ are said to be *positive-linearly dependent* if there are $\lambda_1, \cdots, \lambda_k \geqslant 0$, not all zero, such that

$$\sum_{i=1}^{k} \lambda_i \xi_i = 0.$$

Otherwise, we say that $\xi_1, \cdots, \xi_k$ are *positive-linearly independent*.

Note that the positive-linear independence is much weaker than the usual linear independence. For example, for any non-zero vector $\xi \in \mathbb{R}^n$, and any $\lambda > 0$, ξ and $\lambda\xi$ are linearly dependent; but they are positive-linearly independent. With such a notion, we have the following simple but interesting result.

Proposition 4.11. *Suppose* $f : \mathbb{R}^n \to \mathbb{R}$ *and* $h : \mathbb{R}^n \to \mathbb{R}^\ell$ *are* C^1 *and* $\bar{x}$ *is a local minimum of* $f(\cdot)$ *subject to* $h(x) \leqslant 0$. *Let* $h_x^{j_1}(\bar{x}), \cdots, h_x^{j_k}(\bar{x})$ *be positively linearly independent, where*

$$\{j_1, \cdots, j_k\} = \{j \mid h^j(\bar{x}) = 0\} \equiv J(\bar{x}).$$

Then the Fritz John's multiplier is qualified.

We leave the proof to the readers.

To conclude this section, we state the following result which is due to Han–Mangasarian (1979) whose proof is very similar to that of Theorem 3.3, part (iii).

Theorem 4.12. *Let* $(\bar{x}, \bar{\lambda}^0, \bar{\lambda}, \bar{\mu}) \in G \times [0, 1] \times \mathbb{R}^m \times \mathbb{R}_+^\ell$ *satisfy the Fritz John's necessary condition. Suppose that for every* $z \in \mathbb{R}^n \setminus \{0\}$ *satisfying*

$$g_x(\bar{x})z = 0, \qquad h_x^J(\bar{x})z < 0, \qquad f_x(\bar{x})z \leqslant 0, \qquad (4.18)$$

with $h^J(\cdot)$ *defined in (3.2), the following holds:*

$$z^\top \Lambda_{xx}(\bar{x}, \bar{\lambda}^0, \bar{\lambda}, \bar{\mu})z > 0. \qquad (4.19)$$

Then $\bar{x}$ *is a strictly local minimum of Problem (3.1).*

Exercises

1. Let $f : \mathbb{R}^n \to \mathbb{R}$ and $g : \mathbb{R}^n \to \mathbb{R}^m$. Consider problem

$$\begin{cases} \min \quad f(x), \\ \text{subject to} \quad g(x) = 0. \end{cases}$$

State and prove a theorem similar to Theorem 4.3.

   2. Let $f : \mathbb{R}^n \to \mathbb{R}$ and $h : \mathbb{R}^n \to \mathbb{R}^\ell$. Consider problem

$$\begin{cases} \min & f(x), \\ \text{subject to} & h(x) \leqslant 0. \end{cases}$$

State and prove a theorem similar to Theorem 4.3.

   3. Prove Proposition 4.11.

   4. Let $g(x) = \|x\|^2 - 1$ and $h(x) = a^\top x - b$, where $a \in \mathbb{R}^n$ and $b \in \mathbb{R}$.

   (i) Find conditions for a and b so that the constraints $g(x) = 0$ and $h(x) \leqslant 0$ satisfy the MF qualification condition for any feasible points.

   (ii) Find conditions for a and b so that the constraints $h(x) = 0$ and $g(x) \leqslant 0$ satisfy the MF qualification condition for any feasible points.

   5. Prove Theorem 4.12. (Hint: Mimicking the proof of Theorem 3.1, part (iii).)

Chapter 4

Problems with Convexity and Quasi-Convexity Conditions

For definiteness, in this chapter, $\|\cdot\|$ will be the usual Euclidean norm.

1 Convex Sets and Convex Functions

We first introduce the following definition.

Definition 1.1. (i) A non-empty set $G \subseteq \mathbb{R}^n$ is said to be *convex* if

$$(1 - \lambda)x + \lambda y \in G, \qquad \forall x, y \in G, \quad \lambda \in [0, 1].$$

(ii) For any set $G \subseteq \mathbb{R}^n$, the intersection of all convex sets containing G is called the *convex hull* of G, denoted by $\mathrm{co}(G)$.

It is clear that if $G \subseteq \mathbb{R}^n$ is a subspace, it is convex. But, a convex set might not be a subspace.

Note that convexity is an algebraic property (not a topological property) since only addition and scalar multiplication are involved. We now present some results on convex sets. The following gives a representation of the convex hull $\mathrm{co}(G)$ of G.

Theorem 1.2. (Carathéodory) *Let $G \subseteq \mathbb{R}^n$. Then*

$$\mathrm{co}(G) = \Big\{ \sum_{i=1}^{n+1} \lambda_i x_i \mid x_i \in G, \lambda_i \geqslant 0, \ 1 \leqslant i \leqslant n + 1, \ \sum_{i=1}^{n+1} \lambda_i = 1 \Big\}. \quad (1.1)$$

Proof. Define

$$\begin{cases} K_\ell(G) = \Big\{ \sum_{i=1}^{\ell} \lambda_i x_i \mid x_i \in G, \ \lambda_i \geqslant 0, \ 1 \leqslant i \leqslant \ell, \ \sum_{i=1}^{\ell} \lambda_i = 1 \Big\}, & \ell \geqslant 1, \\ K(G) = \bigcup_{\ell=1}^{\infty} K_\ell(G). \end{cases}$$

Thus, $K(G)$ is the set of all convex combinations of elements in G. Since in the convex combinations, some λ_is are allowed to be zero, we see that

$$K_\ell(G) \subseteq K_{\ell+1}(G), \qquad \forall \ell \geqslant 1.$$

It is easy to check that $K(G)$ is convex, and contains G. Therefore, by the definition of convex hull, $\mathrm{co}(G) \subseteq K(G)$.

On the other hand, we claim that if $C \subseteq \mathbb{R}^n$ is a convex set containing G, then $K(G) \subseteq C$. To show this, we use induction. It is clear that

$$K_1(G) = G \subseteq C.$$

Suppose $K_\ell(G) \subseteq C$. We want to show that $K_{\ell+1}(G) \subseteq C$. In fact, for any

$$y = \sum_{i=1}^{\ell+1} \lambda_i x_i = \sum_{i=1}^{\ell} \lambda_i x_i + \lambda_{\ell+1} x_{\ell+1} \in K_{\ell+1}(G),$$

if $\lambda_{\ell+1} = 1$, then $\lambda_i = 0$, $1 \leqslant i \leqslant \ell$ and

$$y = x_{\ell+1} \in G \subseteq C;$$

and if $\lambda_{\ell+1} = 0$, then $y \in K_\ell(G) \subseteq C$. Now, suppose $0 < \lambda_{\ell+1} < 1$. Then

$$\sum_{i=1}^{\ell} \lambda_i = 1 - \lambda_{\ell+1} > 0, \qquad \sum_{i=1}^{\ell} \frac{\lambda_i}{1 - \lambda_{\ell+1}} = 1,$$

and

$$\sum_{i=1}^{\ell} \frac{\lambda_i}{1 - \lambda_{\ell+1}} x_i \in K_\ell(G) \subseteq C.$$

Hence, by the convexity of C, we have

$$y = (1 - \lambda_{\ell+1}) \Big(\sum_{i=1}^{\ell} \frac{\lambda_i}{1 - \lambda_{\ell+1}} x_i \Big) + \lambda_{\ell+1} x_{\ell+1} \in C.$$

Consequently,

$$K(G) = \bigcup_{\ell=1}^{\infty} K_\ell(G) \subseteq C.$$

Since C is an arbitrary convex set containing G, one has $K(G) \subseteq \mathrm{co}(G)$. Thus, $\mathrm{co}(G) = K(G)$.

Now, we claim that

$$K_\ell(G) = K_{n+1}(G), \qquad \forall \ell > n + 1. \tag{1.2}$$

In fact, let $\ell > n + 1$ and let

$$y = \sum_{i=1}^{\ell} \lambda_i x_i \in K_\ell(G), \qquad \lambda_i \geqslant 0, \quad \sum_{i=1}^{\ell} \lambda_i = 1.$$

Consider $\ell - 1$ vectors $x_1 - x_\ell, x_2 - x_\ell, \cdots, x_{\ell-1} - x_\ell$. Since $\ell - 1 > n$, these $\ell - 1$ vectors are linearly dependent. Hence, there are $\mu_1, \cdots, \mu_{\ell-1} \in \mathbb{R}$, not all zero such that

$$0 = \sum_{i=1}^{\ell-1} \mu_i(x_i - x_\ell) = \sum_{i=1}^{\ell-1} \mu_i x_i - \Big(\sum_{i=1}^{\ell-1} \mu_i\Big) x_\ell = \sum_{i=1}^{\ell} \mu_i x_i,$$

with $\mu_\ell = -\sum_{i=1}^{\ell-1} \mu_i$. Then for any $t \in \mathbb{R}$, we have

$$y = \sum_{i=1}^{\ell} (\lambda_i - t\mu_i) x_i.$$

Let $M = \{i \in \{1, 2, \cdots, \ell\} \mid \mu_i > 0\}$. Since not all μ_i are zero and $\sum_{i=1}^{\ell} \mu_i = 0$, we must have $M \neq \phi$. Therefore, one may let

$$\bar{t} \triangleq \frac{\lambda_{i_0}}{\mu_{i_0}} = \min \Big\{ \frac{\lambda_i}{\mu_i} \mid i \in M \Big\} \geqslant 0.$$

Now, take $t = \bar{t}$. Then for $i \in M$, we have

$$\lambda_i - \bar{t}\mu_i = \mu_i\Big(\frac{\lambda_i}{\mu_i} - \frac{\lambda_{i_0}}{\mu_{i_0}}\Big) \geqslant 0;$$

and for $i \notin M$, $\mu_i \leqslant 0$, so $(\lambda_i - \bar{t}\mu_i) \geqslant 0$. Consequently, we have

$$\lambda_i - \bar{t}\mu_i \geqslant 0, \qquad 1 \leqslant i \leqslant \ell, \quad \lambda_{i_0} - \bar{t}\mu_{i_0} = 0,$$

and

$$\sum_{i=1}^{\ell} (\lambda_i - \bar{t}\mu_i) = \sum_{i=1}^{\ell} \lambda_i - \bar{t} \sum_{i=1}^{\ell} \mu_i = \sum_{i=1}^{\ell} \lambda_i = 1.$$

Therefore, $y \in K_{\ell-1}(G)$, which means that

$$K_\ell(G) = K_{\ell-1}(G), \qquad \forall \ell > n + 1.$$

By induction, we can reach (1.2). Then

$$co(G) = K(G) = \bigcup_{\ell=1}^{\infty} K_\ell(G) = K_{n+1}(G),$$

proving the theorem. $\qquad\qquad\blacksquare$

The interesting point of the above result is that the summation in the right hand side of (1.1) contains at most $(n+1)$ terms. The following proposition collects some other basic properties of convex sets.

Proposition 1.3. (i) *Suppose* $\{G_\lambda, \lambda \in \Lambda\}$ *is a family of convex sets. Then* $\bigcap_{\lambda\in\Lambda} G_\lambda$ *is convex.*

(ii) *If* $A, B \subseteq \mathbb{R}^n$ *are convex, so is*

$$\alpha A + \beta B \equiv \{\alpha x + \beta y \mid x \in A,\ y \in B\}, \qquad \forall \alpha, \beta \in \mathbb{R}. \tag{1.3}$$

In addition, if both A *and* B *are closed and one of them is bounded, then* $\alpha A + \beta B$ *is also closed.*

(iii) *Let* $G_i \subseteq \mathbb{R}^{n_i}$ $(i = 1, 2, \cdots, k)$ *be convex. Then* $G_1 \times \cdots \times G_k$ *is convex in* $\mathbb{R}^{n_1} \times \cdots \times \mathbb{R}^{n_k}$.

(iv) *If* $G \subseteq \mathbb{R}^n$ *is convex, so are* $\bar{G}$ *and* G°.

(v) *If* $G \subseteq \mathbb{R}^n$ *is convex and* $G^\circ \neq \varnothing$, *then*

$$\overline{G^\circ} = \bar{G}, \qquad (\bar{G})^\circ = G^\circ. \tag{1.4}$$

(vi) *The set* G *is convex if and only if* $G = \mathrm{co}(G)$.

(vii) *If* G *is open, so is* $\mathrm{co}(G)$, *and if* G *is compact, so is* $\mathrm{co}(G)$.

Proof. The proofs of (i)–(iii) are left to the readers.

(iv) For any $x, y \in \bar{G}$, there are sequences $x_k, y_k \in G$ such that $x_k \to x$ and $y_k \to y$, as $k \to \infty$ (why?). Then by the convexity of G, we have

$$(1 - \lambda)x_k + \lambda y_k \in G, \qquad \forall k \geqslant 1.$$

Passing to the limit in the above, we obtain

$$(1 - \lambda)x + \lambda y \in \bar{G},$$

which gives the convexity of $\bar{G}$.

Next, we prove that G° is convex. If $G^\circ = \varnothing$, the conclusion is trivial. Suppose now that $G^\circ \neq \varnothing$. Let $x \in G^\circ$ and $y \in \bar{G}$. We show that

$$[x, y) \equiv \{(1 - \lambda)x + \lambda y \mid \lambda \in [0, 1)\} \subseteq G^\circ. \tag{1.5}$$

Fix $\lambda \in (0, 1)$ and we will show that $(1 - \lambda)x + \lambda y \in G^\circ$. To this end, we first find a $\delta > 0$ such that $B_{2\delta}(x) \subseteq G^\circ$. Since $y \in \bar{G}$, there exists a $\tilde{y} \in G$ such that

$$\|y - \tilde{y}\| < \frac{(1 - \lambda)\delta}{\lambda}.$$

Define

$$z = x - \frac{\lambda}{1-\lambda}(\widetilde{y} - y). \tag{1.6}$$

Then

$$\|z - x\| = \frac{\lambda}{1-\lambda}\|\widetilde{y} - y\| < \delta,$$

which means $z \in B_\delta(x) \subseteq B_{2\delta}(x) \subseteq G^\circ$. Further,

$$\begin{aligned}(1-\lambda)x + \lambda y &= (1-\lambda)\Big[z + \frac{\lambda}{1-\lambda}(\widetilde{y} - y)\Big] + \lambda y \\ &= (1-\lambda)z + \lambda(\widetilde{y} - y) + \lambda y = (1-\lambda)z + \lambda\widetilde{y}.\end{aligned} \tag{1.7}$$

We claim that

$$B_{(1-\lambda)\delta}\big((1-\lambda)z + \lambda\widetilde{y}\big) \subseteq G.$$

In fact, for any $\zeta \in B_{(1-\lambda)\delta}\big((1-\lambda)z + \lambda\widetilde{y}\big)$, let

$$\eta = z + \frac{\zeta - [(1-\lambda)z + \lambda\widetilde{y}]}{1-\lambda}.$$

Since

$$\|\eta - z\| = \frac{\|\zeta - [(1-\lambda)z + \lambda\widetilde{y}]\|}{1-\lambda} < \frac{(1-\lambda)\delta}{1-\lambda} = \delta,$$

one has

$$\|\eta - x\| \leqslant \|\eta - z\| + \|z - x\| < 2\delta,$$

which leads to $\eta \in B_{2\delta}(x) \subseteq G^\circ \subseteq G$. Then by the convexity of G,

$$\zeta = (1-\lambda)(\eta - z) + (1-\lambda)z + \lambda\widetilde{y} = (1-\lambda)\eta + \lambda\widetilde{y} \in G.$$

Hence, noting (1.7), and by the choice of ζ, one has

$$B_{(1-\lambda)\delta}\big((1-\lambda)x + \lambda y\big) = B_{(1-\lambda)\delta}\big((1-\lambda)z + \lambda\widetilde{y}\big) \subseteq G.$$

This proves (1.5) for $x \in G^\circ$, $y \in \bar{G}$, and $\lambda \in (0,1)$. Then

$$(1-\lambda)x + \lambda y \in G^\circ, \qquad \forall x, y \in G^\circ, \ \lambda \in [0,1],$$

which gives the convexity of G°.

(v) First of all, since $G^\circ \subseteq G$, we have

$$\overline{G^\circ} \subseteq \bar{G}.$$

Next, let $y \in \bar{G}$. Then (1.5) tells us that for any $x \in G^\circ$, the segment $[x, y) \subseteq G^\circ$, which implies $y \in \overline{G^\circ}$. Hence,

$$\bar{G} \subseteq \overline{G^\circ},$$

proving the first relation in (1.4).

Next, since $G \subseteq \bar{G}$, we have

$$G^\circ \subseteq (\bar{G})^\circ.$$

On the other hand, let $x \in (\bar{G})^\circ$. Then for some $\delta > 0$, $B_{2\delta}(x) \subseteq (\bar{G})^\circ \subseteq \bar{G}$. Let $y \in G^\circ$ and choose $\lambda \in (0, 1)$ such that

$$\frac{(1 - \lambda)\|x - y\|}{\lambda} < \delta.$$

Define

$$z = \frac{x - (1 - \lambda)y}{\lambda}.$$

Then

$$\|z - x\| = \left\| \frac{x - (1 - \lambda)y}{\lambda} - x \right\| = \frac{1 - \lambda}{\lambda}\|x - y\| < \delta,$$

which means $z \in B_\delta(x) \subseteq \bar{G}$. Hence, by (1.5), we have $[y, z) \subseteq G^\circ$. Hence,

$$x = \lambda z + (1 - \lambda)y \in G^\circ,$$

which yields

$$(\bar{G})^\circ \subseteq G^\circ.$$

Therefore, the second relation in (1.4) follows.

(vi) If $G = \text{co}(G)$ then, of course, G is convex. Conversely, if G is convex, then $K(G) \subseteq G \subseteq \text{co}(G) = K(G)$, leading to $G = \text{co}(G)$.

(vii) Since $G \subseteq \text{co}(G)$, and G is open and non-empty, we have $(\text{co}(G))^\circ \neq \phi$. Then by (iv), $(\text{co}(G))^\circ$ is convex, and

$$G = G^\circ \subseteq (\text{co}(G))^\circ.$$

Consequently, by the definition of $\text{co}(G)$,

$$\text{co}(G) \subseteq (\text{co}(G))^\circ \subseteq \text{co}(G).$$

Hence, $\text{co}(G) = (\text{co}(G))^\circ$ which is open.

Next, if we denote

$$P_{n+1} = \Big\{ (\lambda_1, \cdots, \lambda_{n+1}) \in \mathbb{R}^{n+1} \mid \lambda_i \geqslant 0, \ \sum_{i=1}^{n+1} \lambda_i = 1 \Big\},$$

then P_{n+1} is compact. Therefore, when G is compact, $P_{n+1} \times G^{n+1}$ is also compact. Also, if we denote

$$\varphi(\lambda, x) = \sum_{i=1}^{n+1} \lambda_i x_i, \qquad \forall (\lambda, x) \in P_{n+1} \times (\mathbb{R}^n)^{n+1},$$

then $\varphi(\cdot\,,\cdot)$ is continuous and

$$\mathrm{co}(G) = K_{n+1}(G) = \varphi(P_{n+1} \times G^{n+1}),$$

which must be compact (by Theorem 3.3 of Chapter 1). $\square$

Note that if G is just closed, $\mathrm{co}(G)$ is not necessarily closed. Here is a simple example. Let

$$G = \{(x,y) \in \mathbb{R}^2 \mid y > 0, \ |x|y \geqslant 1\}.$$

Then G is closed. But

$$\mathrm{co}(G) = \{(x,y) \in \mathbb{R}^2 \mid x > 0\},$$

which is open.

Now, let $G \subseteq \mathbb{R}^n$ be a non-empty closed convex set. We look at the following minimization problem: For given $x \in \mathbb{R}^n$,

$$\begin{cases} \min \|x - y\|, \\ \text{subject to } y \in G. \end{cases}$$

For such a problem, we have the following proposition which will lead to a fundamental result for convex sets.

Proposition 1.4. *Let $G \subseteq \mathbb{R}^n$ be a non-empty convex and closed set. Then for any $x \in \mathbb{R}^n$, there exists a unique $\bar{x} \in G$ such that*

$$\|x - \bar{x}\| = \inf_{y \in G} \|x - y\| \triangleq d(x, G). \tag{1.8}$$

Moreover, $\bar{x}$ is characterized by the following:

$$(x - \bar{x})^\top (y - \bar{x}) \leqslant 0, \qquad \forall y \in G. \tag{1.9}$$

Proof. For fixed $x \in \mathbb{R}^n$, we can find a sequence $x_k \in G$ such that

$$\lim_{k \to \infty} \|x_k - x\| = \inf_{y \in G} \|x - y\| \equiv d.$$

Note that for any $a, b \in \mathbb{R}^n$,

$$\|a + b\|^2 + \|a - b\|^2 = 2\|a\|^2 + 2\|b\|^2,$$

which is called the *parallelogram law*. Thus, taking $a = x - x_k$ and $b = x - x_\ell$, we have

$$\|x_k - x_\ell\|^2 = 2\|x - x_k\|^2 + 2\|x - x_\ell\|^2 - 4\left\|x - \frac{x_k + x_\ell}{2}\right\|^2$$
$$\leqslant 2\|x - x_k\|^2 + 2\|x - x_\ell\|^2 - 4d^2,$$

since by the convexity of G, $\frac{x_k + x_\ell}{2} \in G$. Then we see that $\{x_k\}$ is a Cauchy sequence. Hence, by the closeness of G, we see that $x_k \to \bar{x} \in G$, as $k \to \infty$. This gives the existence of $\bar{x}$ satisfying (1.8). Suppose there is another one $\bar{y} \in G$ such that

$$\|x - \bar{y}\| = d.$$

Then similar to the above, we have

$$\begin{aligned}
\|\bar{x} - \bar{y}\|^2 &= 2\|x - \bar{x}\|^2 + 2\|x - \bar{y}\|^2 - 4\left\|x - \tfrac{\bar{x}+\bar{y}}{2}\right\|^2 \\
&\leqslant 2\|x - \bar{x}\|^2 + 2\|x - \bar{y}\|^2 - 4d^2 = 0.
\end{aligned}$$

Hence, $\bar{y} = \bar{x}$, proving the uniqueness.

Finally, for any $y \in G$ and $\alpha \in (0,1)$, we have

$$\bar{x} + \alpha(y - \bar{x}) = (1 - \alpha)\bar{x} + \alpha y \in G.$$

Thus,

$$\begin{aligned}
\|x - \bar{x}\|^2 = d^2 &\leqslant \|x - [\bar{x} + \alpha(y - \bar{x})]\|^2 \\
&\leqslant \|x - \bar{x}\|^2 - 2\alpha(x - \bar{x})^\top(y - \bar{x}) + \alpha^2\|y - \bar{x}\|^2.
\end{aligned}$$

Then

$$(x - \bar{x})^\top(y - \bar{x}) \leqslant \frac{\alpha}{2}\|y - \bar{x}\|^2 \to 0, \qquad \text{as } \alpha \to 0,$$

proving (1.9). Conversely, if (1.9) holds, then for any $y \in G$, we have

$$\begin{aligned}
\|x - y\|^2 &= \|x - \bar{x} - (y - \bar{x})\|^2 \\
&= \|x - \bar{x}\|^2 - 2(x - \bar{x})^\top(y - \bar{x}) + \|y - \bar{x}\|^2 \geqslant \|x - \bar{x}\|^2.
\end{aligned}$$

Hence, $\bar{x}$ reaches the minimum of $y \mapsto \|x - y\|$ over $y \in G$. $\qquad\square$

In the above, we call $d(x, G)$ the distance between x and G. We usually refer to (1.9) as a *variational inequality*. Note that in the above if $x \notin G$, then by letting $a = x - \bar{x} \neq 0$, (1.9) means

$$a^\top(x - \bar{x}) = \|x - \bar{x}\|^2 > 0 \geqslant (x - \bar{x})^\top(y - \bar{x}) = a^\top(y - \bar{x}), \quad \forall y \in G. \quad (1.10)$$

Therefore, we have

$$\begin{cases} x \in \{z \in \mathbb{R}^n \mid a^\top z > a^\top \bar{x} > 0\}, \\ G \subseteq \{y \in \mathbb{R}^n \mid a^\top y \leqslant a^\top \bar{x}\}. \end{cases}$$

This tells us that the hyperplane $\{y \in \mathbb{R}^n \mid a^\top y = a^\top \bar{x}\}$ *separates* the point x from the convex set G. Clearly, (1.10) is also equivalent to the following:

$$a^\top x \geqslant \|a\|^2 + a^\top \bar{x} \geqslant \|a\|^2 + a^\top y > a^\top y, \qquad \forall y \in G.$$

We now extend the above to the following which is called the *Separation Theorem* of convex sets.

Theorem 1.5. *Let $G_1, G_2 \subseteq \mathbb{R}^n$ be two convex sets such that*

$$G_1^\circ \bigcap G_2 = \varnothing. \qquad (1.11)$$

Then there exists a vector $a \in \mathbb{R}^n$ with $\|a\| = 1$ such that

$$a^\top x_1 \leqslant a^\top x_2, \qquad \forall x_1 \in G_1, \ x_2 \in G_2. \qquad (1.12)$$

Further, if

$$\overline{G}_1 \bigcap \overline{G}_2 = \varnothing, \qquad (1.13)$$

and either G_1 or G_2 is bounded, then there exists a constant $c \in \mathbb{R}$, such that (1.12) is replaced by

$$a^\top x_1 < c < a^\top x_2, \qquad \forall x_1 \in G_1, \ x_2 \in G_2. \qquad (1.14)$$

Proof. First, let (1.13) hold and, say, G_1 is bounded. Then by Proposition 1.3, one has that

$$\overline{G}_2 - \overline{G}_1 \equiv \{a_2 - a_1 \mid a_i \in \overline{G}_i, \ i = 1, 2\} \subseteq \mathbb{R}^n$$

is convex and closed, and $0 \notin \overline{G}_2 - \overline{G}_1$. Hence, by Proposition 1.4, there exists an $\bar{x} \in \overline{G}_2 - \overline{G}_1$ such that

$$0 < \|\bar{x}\| = \inf_{y \in \overline{G}_2 - \overline{G}_1} \|y\|.$$

Let $a = \frac{\bar{x}}{\|\bar{x}\|}$. Then $\|a\| = 1$ and

$$a^\top(y - \bar{x}) = -\frac{1}{\|\bar{x}\|}(0 - \bar{x})^\top(y - \bar{x}) \geqslant 0, \qquad \forall y \in \overline{G}_2 - \overline{G}_1.$$

Therefore,

$$a^\top y \geqslant a^\top \bar{x} = \|\bar{x}\|, \qquad \forall y \in \overline{G}_2 - \overline{G}_1.$$

Then by the definition of $\overline{G}_2 - \overline{G}_1$, we obtain that

$$a^\top x_2 \geqslant a^\top x_1 + \|\bar{x}\|, \qquad \forall x_1 \in \overline{G}_1, \ x_2 \in \overline{G}_2.$$

Hence, (1.14) follows by setting

$$c = \frac{\|\bar{x}\|}{2} + \sup_{x_1 \in \overline{G}_1} a^\top x_1.$$

Next, we let (1.11) hold, but (1.13) fails. Let $G = G_2 - G_1$. Then G is convex (not necessarily closed), and

$$0 \in \partial G.$$

Hence, we can find a sequence $y_k \in \mathbb{R}^n \setminus \overline{G}$ ($k \geqslant 1$) such that

$$\lim_{k \to \infty} \|y_k\| = 0.$$

Now, for each $k \geqslant 1$, since $y_k \notin \overline{G}$, applying what we have proved (with G_1 being $\{y_k\}$ and G_2 being G), we can find a $c_k > 0$, and an $a_k \in \mathbb{R}^m$ with $\|a_k\| = 1$, such that

$$a_k^\top y_k < c_k < a_k^\top x, \qquad \forall x \in G.$$

Since $\|a_k\| = 1$, we may assume that $a_k \to a$ for some $\|a\| = 1$. Then passing to the limit in the above, we obtain

$$0 \leqslant a^\top x, \qquad \forall x \in G.$$

Consequently, by the definition of G, we obtain (1.12). ∎

We point out that (1.13) alone (without assuming the boundedness of either G_1 or G_2) does not imply (1.14). Here is a simple example. Let

$$G_1 = \{(x, 0) \in \mathbb{R}^2 \mid x \in \mathbb{R}\}, \qquad G_2 = \{(x, y) \in \mathbb{R}^2 \mid xy \geqslant 1\}.$$

Then both G_1 and G_2 are convex and closed. Moreover, they are disjoint. However, (1.14) fails.

We now introduce the following notions.

Definition 1.6. Let $G \subseteq \mathbb{R}^n$ be convex, and $f : G \to \mathbb{R}$.

(i) $f(\cdot)$ is said to be *convex* if

$$f(\lambda x + (1 - \lambda)y) \leqslant \lambda f(x) + (1 - \lambda)f(y), \quad \forall x, y \in G, \ \lambda \in (0, 1). \quad (1.15)$$

Further, if in the above, a strict inequality holds for any $x \neq y$, we say that $f(\cdot)$ is *strictly convex*.

(ii) $f(\cdot)$ is said to be (*strictly*) *concave* if $-f(\cdot)$ is (strictly) convex.

From the above (ii), we see that having discussed convex functions, the discussion of concave functions is not necessary. Let us present some simple examples of convex functions.

Example 1.7. Let $f(x) = |x|$ and $g(x) = x^2$. We show that they are convex. In fact, for any $x, y \in \mathbb{R}$ and $\lambda \in (0, 1)$,

$$|\lambda x + (1 - \lambda)y| \leqslant \lambda|x| + (1 - \lambda)|y|,$$

proving the convexity of $f(x)$. On the other hand,

$$\left(\lambda x + (1 - \lambda)y\right)^2 = \lambda^2 x^2 + 2\lambda(1 - \lambda)xy + (1 - \lambda)y^2$$
$$\leqslant \lambda^2 x^2 + \lambda(1 - \lambda)(x^2 + y^2) + (1 - \lambda)y^2 = \lambda x^2 + (1 - \lambda)y^2,$$

proving the convexity of $g(x)$.

We now present the following result concerning convex functions.

Proposition 1.8. *Let $G \subseteq \mathbb{R}^n$ be convex and $f : G \to \mathbb{R}$.*

(i) $f(\cdot)$ *is convex if and only if the epigraph of $f(\cdot)$ defined by the following is convex:*

$$\text{epi}(f) = \{x, y) \in G \times \mathbb{R} \mid f(x) \leqslant y\}. \tag{1.16}$$

(ii) **(Jensen's inequality)** $f(\cdot)$ *is convex if and only if for any finitely many $\lambda_1, \cdots, \lambda_k \geqslant 0$ with $\sum_{i=1}^{k} \lambda_i = 1$ and $x_1, \cdots, x_k \in G$,*

$$f\left(\sum_{i=1}^{k} \lambda_i x_i\right) \leqslant \sum_{i=1}^{k} \lambda_i f(x_i). \tag{1.17}$$

Proof. Suppose $\text{epi}(f)$ is convex. Let $x_1, x_2 \in G$. Then

$$(x_1, f(x_1)), (x_2, f(x_2)) \in \text{epi}(f).$$

By the convexity of $\text{epi}(f)$, for any $\lambda \in (0, 1)$, we have

$$\lambda(x_1, f(x_1)) + (1 - \lambda)(x_2, f(x_2))$$
$$= (\lambda x_1 + (1 - \lambda)x_2, \lambda f(x_1) + (1 - \lambda)f(x_2)) \in \text{epi}(f),$$

which means that

$$f(\lambda x_1 + (1 - \lambda)x_2) \leqslant \lambda f(x_1) + (1 - \lambda)f(x_2),$$

giving the convexity of $f(\cdot)$. Conversely, if $f(\cdot)$ is convex, then for any $(x_1, y_1), (x_2, y_2) \in \text{epi}(f)$, and $\lambda \in (0, 1)$, if we denote

$$(x_3, y_3) = \lambda(x_1, y_1) + (1 - \lambda)(x_2, y_2) = (\lambda x_1 + (1 - \lambda)x_2, \lambda y_1 + (1 - \lambda)y_2),$$

then by the convexity of $f(\cdot)$, we have

$$f(x_3) = f(\lambda x_1 + (1 - \lambda)x_2) \leqslant \lambda f(x_1) + (1 - \lambda)f(x_2)$$
$$\leqslant \lambda y_1 + (1 - \lambda)y_2 = y_3,$$

which means $(x_3, y_3) \in \text{epi}(f)$. Hence, $\text{epi}(f)$ is convex.

(ii) The proof is left to the readers. □

Next, we would like to look at the continuity of convex functions. The interesting point here is that the definition of convex function only involves the convexity of the domain G (which is a linear structure property) and the order on $\mathbb{R}$. Thus, the continuity of convex function shows the perfect compatibility of the linear structure, the topological structure of $\mathbb{R}^n$, and the order of $\mathbb{R}$.

Theorem 1.9. *Let $G \subseteq \mathbb{R}^n$ be convex with $G^\circ \neq \varnothing$, and $f : G \to \mathbb{R}$ be convex. Then $f(\cdot)$ is locally Lipschitz continuous in G°, i.e., for any $\bar{x} \in G$, there exists an $M(\bar{x}) > 0$ and a $\delta > 0$ with $B_\delta(\bar{x}) \subseteq G$ such that*

$$|f(x) - f(y)| \leqslant M(\bar{x})\|x - y\|, \qquad \forall x, y \in B_\delta(\bar{x}).$$

Proof. First, we prove that $f(\cdot)$ is locally bounded from above, i.e., for any $\bar{x} \in G^\circ$, there exists a neighborhood of $\bar{x}$ on which $f(\cdot)$ is bounded from above. To this end, we let $\varepsilon > 0$ be small enough such that the closed cube $Q_\varepsilon(\bar{x})$ centered at $\bar{x}$ with side length 2ε is contained in G. Let $K = \{a_k \mid 1 \leqslant k \leqslant 2^n\}$ be the set of all extreme points of $Q_\varepsilon(\bar{x})$, i.e., each a_k has the following property: The components of $a_k - \bar{x}$ are $\pm\varepsilon$ only. Clearly,

$$Q_\varepsilon(\bar{x}) = \text{co}(K) = \Big\{ \sum_{k=1}^{2^n} \lambda_k a_k \mid \lambda_k \geqslant 0, \ \sum_{k=1}^{2^n} \lambda_k = 1 \Big\}.$$

Then for any $y \in Q_\varepsilon(\bar{x})$, let

$$y = \sum_{k=1}^{2^n} \lambda_k a_k, \qquad \text{for some } \lambda_k \geqslant 0, \ \sum_{k=1}^{2^n} \lambda_k = 1.$$

By Jensen's inequality, we have

$$f(y) = f\Big(\sum_{k=1}^{2^n} \lambda_k a_k \Big) \leqslant \sum_{k=1}^{2^n} \lambda_k f(a_k) \leqslant \max\{f(a_k) \mid 1 \leqslant k \leqslant 2^n\} = \bar{M}_\varepsilon(\bar{x}).$$

This proves the local boundedness of $f(\cdot)$ from above.

Next, we show that $f(\cdot)$ is locally bounded from below. To this end, suppose $f(\cdot)$ is bounded from above in $Q_\varepsilon(\bar{x})$, but not bounded from below. Then there will be a sequence $x_k \in Q_\varepsilon(\bar{x})$ such that $f(x_k) \to -\infty$. Since $Q_\varepsilon(\bar{x})$ is compact, we may assume that $x_k \to y \in Q_\varepsilon(\bar{x})$. We pick a $z \in Q_\varepsilon(\bar{x})$ such that for some $\delta > 0$, $x_k \notin B_\delta(z) \subseteq Q_\varepsilon(\bar{x})$ for all $k \geqslant 1$. Now, we define

$$z_k = z + \frac{\delta}{\|z - x_k\|}(z - x_k), \qquad k \geqslant 1.$$

Then $z_k \in B_\delta(z) \subseteq Q_\varepsilon(\bar{x})$, which implies $f(z_k) \leqslant \bar{M}_\varepsilon(\bar{x})$, and

$$z = \frac{\delta}{\delta + \|z - x_k\|} x_k + \frac{\|z - x_k\|}{\delta + \|z - x_k\|} z_k.$$

Hence,

$$f(z) \leqslant \frac{\delta}{\delta + \|z - x_k\|} f(x_k) + \frac{\|z - x_k\|}{\delta + \|z - x_k\|} f(z_k)$$

$$\leqslant \frac{\delta}{\delta + \|z - x_k\|} f(x_k) + \bar{M}_\varepsilon(\bar{x}).$$

This leads to a contradiction since the right hand side will go to $-\infty$ as $k \to \infty$. Hence, we have the local boundedness of $f(\cdot)$.

We now let $\bar{x} \in G^\circ$ and $\delta > 0$ such that $\bar{B}_{4\delta}(\bar{x}) \subseteq G$ and let

$$M_{4\delta}(\bar{x}) = \sup_{x \in B_{4\delta}(\bar{x})} |f(x)| < \infty.$$

For any $x, y \in G \cap B_\delta(\bar{x})$, and $x \neq y$, let $\xi = \frac{y-x}{\|y-x\|}$. Then $\|\xi\| = 1$. Consider the following equation for t:

$$0 = \|x + t\xi - \bar{x}\|^2 - 16\delta^2 = t^2 - 2\xi^\top(\bar{x} - x)t + \|x - \bar{x}\|^2 - 16\delta^2,$$

which has two roots:

$$t = \frac{2\xi^\top(\bar{x} - x) \pm \sqrt{4[\xi^\top(x - \bar{x})]^2 + 4[16\delta^2 - \|x - \bar{x}\|^2]}}{2}$$

$$= \xi^\top(\bar{x} - x) \pm \sqrt{[\xi^\top(x - \bar{x})]^2 + 16\delta^2 - \|x - \bar{x}\|^2}.$$

Thus, if we let (note $\|x - \bar{x}\| < \delta$)

$$\begin{cases} -\alpha_{4\delta} = \xi^\top(\bar{x} - x) - \sqrt{[\xi^\top(x - \bar{x})]^2 + 16\delta^2 - \|x - \bar{x}\|^2} < 0, \\ \beta_{4\delta} = \xi^\top(\bar{x} - x) + \sqrt{[\xi^\top(x - \bar{x})]^2 + 16\delta^2 - \|x - \bar{x}\|^2} > 0, \end{cases}$$

then by the convexity of $t \mapsto \|x + t\xi - \bar{x}\|^2 - 16\delta^2$, one has

$$\|x + t\xi - \bar{x}\|^2 - 16\delta^2 < 0, \qquad \forall t \in (-\alpha_{4\delta}, \beta_{4\delta}),$$

i.e.,

$$x + t\xi \in \bar{B}_{4\delta}(\bar{x}), \qquad \forall t \in [-\alpha_{4\delta}, \beta_{4\delta}].$$

Likewise, if we replace 4δ by 3δ in the above, we may define

$$\begin{cases} -\alpha_{3\delta} = \xi^\top(\bar{x} - x) - \sqrt{[\xi^\top(x - \bar{x})]^2 + 9\delta^2 - \|x - \bar{x}\|^2} < 0, \\ \beta_{3\delta} = \xi^\top(\bar{x} - x) + \sqrt{[\xi^\top(x - \bar{x})]^2 + 9\delta^2 - \|x - \bar{x}\|^2} > 0, \end{cases}$$

and

$$x + t\xi \in \bar{B}_{3\delta}(\bar{x}), \qquad t \in [-\alpha_{3\delta}, \beta_{3\delta}].$$

Further, noting $|\xi^\top(x - \bar{x})| \leqslant \|x - \bar{x}\|$, we have

$$\beta_{4\delta} - \beta_{3\delta} = \alpha_{4\delta} - \alpha_{3\delta}$$

$$= \sqrt{[\xi^\top(x - \bar{x})]^2 + 16\delta^2 - \|x - \bar{x}\|^2} - \sqrt{[\xi^\top(x - \bar{x})]^2 + 9\delta^2 - \|x - \bar{x}\|^2}$$

$$= \frac{7\delta^2}{\sqrt{[\xi^\top(x - \bar{x})]^2 + 16\delta^2 - \|x - \bar{x}\|^2} + \sqrt{[\xi^T(x - \bar{x})]^2 + 9\delta^2 - \|x - \bar{x}\|^2}}$$

$$\geqslant \frac{7\delta^2}{7\delta} = \delta.$$

Now, we define

$$\theta(t) = f(x + t\xi), \qquad t \in [-\alpha_{4\delta}, \beta_{4\delta}].$$

Then for any $t, s \in [-\alpha_{4\delta}, \beta_{4\delta}]$, and $\lambda \in (0, 1)$, we have

$$\theta(\lambda t + (1 - \lambda)s) = f\big(\lambda(x + t\xi) + (1 - \lambda)(x + s\xi)\big)$$
$$\leqslant \lambda f(x + t\xi) + (1 - \lambda)f(x + s\xi) = \lambda\theta(t) + (1 - \lambda)\theta(s).$$

Thus, $\theta(\cdot)$ is convex. Since for $-\alpha_{4\delta} \leqslant t_1 < t_2 < t_3 \leqslant \beta_{4\delta}$, one has

$$t_2 = \frac{t_3 - t_2}{t_3 - t_1}t_1 + \frac{t_2 - t_1}{t_3 - t_1}t_3.$$

Thus, taking $\lambda = \frac{t_3 - t_2}{t_3 - t_1}$, we have

$$\theta(t_2) = \theta(\lambda t_1 + (1 - \lambda)t_3) \leqslant \lambda\theta(t_1) + (1 - \lambda)\theta(t_3)$$
$$= \frac{t_3 - t_2}{t_3 - t_1}\theta(t_1) + \frac{t_2 - t_1}{t_3 - t_1}\theta(t_3).$$

This implies

$$\frac{\theta(t_2) - \theta(t_1)}{t_2 - t_1} \leqslant \frac{\theta(t_3) - \theta(t_1)}{t_3 - t_1},$$

and

$$\frac{\theta(t_3) - \theta(t_2)}{t_3 - t_2} \geqslant \frac{\theta(t_3) - \theta(t_1)}{t_3 - t_1}.$$

Combining the above, we see that

$$\frac{\theta(t_2) - \theta(t_1)}{t_2 - t_1} \leqslant \frac{\theta(t_3) - \theta(t_2)}{t_3 - t_2}, \qquad -\alpha_{4\delta} \leqslant t_1 < t_2 < t_3 \leqslant \beta_{4\delta}. \qquad (1.18)$$

We claim that

$$\beta_{3\delta} = \xi^\top(\bar{x} - x) + \sqrt{[\xi^\top(x - \bar{x})]^2 + 9\delta^2 - \|x - \bar{x}\|^2} > 2\delta,$$

which is equivalent to

$$[\xi^\top(x - \bar{x})]^2 + 9\delta^2 - \|x - \bar{x}\|^2 > \left(2\delta - \xi^\top(\bar{x} - x)\right)^2$$
$$= 4\delta^2 - 4\delta\xi^\top(\bar{x} - x) + [\xi^\top(\bar{x} - x)]^2.$$

This is equivalent to

$$5\delta^2 > \|x - \bar{x}\|^2 + 4\delta\xi^\top(x - \bar{x}),$$

which is true since $\|x - \bar{x}\| < \delta$ and $\|\xi\| = 1$. Now, applying (1.18) to $0 < \|y - x\| < 2\delta < \beta_{3\delta} < \beta_{4\delta}$, we have

$$\frac{f(y) - f(x)}{\|y - x\|} = \frac{\theta(\|y - x\|) - \theta(0)}{\|y - x\|} \leqslant \frac{\theta(\beta_{4\delta}) - \theta(\beta_{3\delta})}{\beta_{4\delta} - \beta_{3\delta}}$$
$$= \frac{f(x + \beta_{4\delta}\xi) - f(x + \beta_{3\delta}\xi)}{\beta_{4\delta} - \beta_{3\delta}} \leqslant \frac{2M_{4\delta}(\bar{x})}{\delta}.$$

Also, applying (1.18) to $-\alpha_{4\delta} < -\alpha_{3\delta} < 0 < \|y - x\|$, one gets

$$\frac{f(y) - f(x)}{\|y - x\|} = \frac{\theta(\|y - x\|) - \theta(0)}{\|y - x\|} \geqslant \frac{\theta(-\alpha_{3\delta}) - \theta(-\alpha_{4\delta})}{\alpha_{4\delta} - \alpha_{3\delta}}$$
$$= \frac{f(x - \alpha_{3\delta}\xi) - f(x - \alpha_{4\delta}\xi)}{\alpha_{4\delta} - \alpha_{3\delta}} \geqslant -\frac{2M_{4\delta}(\bar{x})}{\delta}.$$

Hence,

$$|f(y) - f(x)| \leqslant \frac{2M_{4\delta}(\bar{x})}{\delta}\|y - x\|, \qquad \forall x, y \in B_\delta(\bar{x}).$$

This proves the local Lipschitz continuity of $f(\cdot)$. $\qquad\square$

Next, we would like to look at the case that $f(\cdot)$ is differentiable.

Proposition 1.10. *Suppose $G \subseteq \mathbb{R}^n$ is convex with $G^\circ \neq \varnothing$.*

(i) *Let $f : G \to \mathbb{R}$ be C^1. Then $f(\cdot)$ is convex if and only if*

$$f(y) \geqslant f(x) + f_x(x)(y - x), \qquad \forall x, y \in G. \tag{1.19}$$

(ii) *Let $f : G \to \mathbb{R}$ be C^2. Then $f(\cdot)$ is convex if and only if the Hessian matrix $f_{xx}(\cdot)$ of $f(\cdot)$ is positive semi-definite on G.*

Proof. (i) Suppose $f(\cdot)$ is convex. Then for any $\varepsilon \in (0, 1]$,

$$\frac{f(x + \varepsilon(y - x)) - f(x)}{\varepsilon} = \frac{f((1 - \varepsilon)x + \varepsilon y) - f(x)}{\varepsilon}$$
$$\leqslant \frac{(1 - \varepsilon)f(x) + \varepsilon f(y) - f(x)}{\varepsilon} \leqslant f(y) - f(x),$$

which implies (by letting $\varepsilon \to 0$)

$$f_x(x)(y - x) \leqslant f(y) - f(x),$$

proving (1.19).

Conversely, for any $x, y \in G$ and $\lambda \in [0, 1]$, let

$$z = \lambda x + (1 - \lambda)y.$$

By (1.19) with x replaced by z,

$$f(x) \geqslant f(z) + f_x(z)(x - z),$$
$$f(y) \geqslant f(z) + f_x(z)(y - z).$$

Hence,

$$\lambda f(x) + (1 - \lambda)f(y) \geqslant f(z) + f_x(z)[\lambda x + (1 - \lambda)y - z]$$
$$= f(z) = f(\lambda x + (1 - \lambda)y),$$

proving the convexity of $f(\cdot)$.

(ii) Suppose $f_{xx}(\cdot)$ is positive semi-definite on G. Then, by Taylor expansion, we have

$$f(y) = f(x) + f_x(x)(y - x) + \frac{1}{2}(y - x)^{\top}f_{xx}(x + \theta(y - x))(y - x) \qquad (1.20)$$
$$\geqslant f(x) + f_x(x)(y - x), \qquad \forall x, y \in G.$$

By (i), (1.20) implies the convexity of $f(\cdot)$.

Conversely, suppose $f_{xx}(\cdot)$ is not positive semi-definite at some point in G. Since $f(\cdot)$ is C^2, we may assume that for some interior point x, and some $y \in G$,

$$(y - x)^{\top}f_{xx}(x)(y - x) < 0.$$

Then by continuity again (choosing y sufficiently close to x), we can assume that

$$(y - x)^{\top}f_{xx}(x + \theta(y - x))(y - x) < 0.$$

Hence, for such x and y, the inequality in (1.20) will not hold and therefore, $f(\cdot)$ will not be convex, a contradiction. $\qquad \square$

Proposition 1.11. *Let $f : G \to \mathbb{R}$ be convex and $g : f(G) \to \mathbb{R}$ be convex and non-decreasing. Then $g \circ f : G \to \mathbb{R}$ is convex.*

Proof. For any $x, y \in G$ and $\lambda \in (0, 1)$,

$$g(f(\lambda x + (1 - \lambda)y)) \leqslant g(\lambda f(x) + (1 - \lambda)f(y))$$
$$\leqslant \lambda g(f(x)) + (1 - \lambda)g(f(y)),$$

which gives the convexity of $g \circ f$. $\square$

Exercises

1. Let $x \in \mathbb{R}^n$ and $r > 0$. Show that the set $B_r(x) = \{x \in \mathbb{R}^n \mid \|x\| < r\}$ is convex.

2. Let $a, b \in \mathbb{R}^n$. Show that the set $[a, b] = \{x \in \mathbb{R}^n \mid x = \lambda a + (1 - \lambda)b, \lambda \in [0, 1]\}$ is convex.

3. Let $a \in \mathbb{R}^n$ and $b \in \mathbb{R}$. Show that the set $\{x \in \mathbb{R}^n \mid a^\top x \geqslant b\}$ is convex.

4. Prove (i)–(iii) of Proposition 1.3.

5. Present an example that G_1 and G_2 are convex and closed, but $\alpha G_1 + \beta G_2$ is not closed for some $\alpha, \beta \in \mathbb{R}$.

6. Find the examples of convex and non-empty sets G_1 and G_2 so that the following will be true, respectively:

(i) $G_1 \cup G_2$ is convex;

(ii) $G_1 \cup G_2$ is not convex;

(iii) $G_1 \setminus G_2$ is convex;

(iv) $G_1 \setminus G_2$ is not convex;

7. Let $f(x) = (x - 1)^2 - 2x^2 + (ax + b)^2$ be a function of one variable. Find $a, b \in \mathbb{R}$ such that $f(x)$ is convex.

8. Let $\mathbf{1} = (1, 1, \cdots, 1)^\top \in \mathbb{R}^n$ and $f(x) = 1 - \cos(\mathbf{1}^\top x)$. Find an $r > 0$ such that $f(x)$ is convex on $B_r(0)$. What is the maximum possible value of r?

9. Find an example that $f, g : \mathbb{R} \to \mathbb{R}$ both are convex, but $g \circ f : \mathbb{R} \to \mathbb{R}$ is not convex.

10. Find an example that $f : \mathbb{R} \to \mathbb{R}$ is convex, and $g : \mathbb{R} \to \mathbb{R}$ is increasing such that $g \circ f : \mathbb{R} \to \mathbb{R}$ is not convex.

11. Let $f : \mathbb{R}^n \to \mathbb{R}$ be convex. Show that for any $\alpha \in \mathbb{R}$, the set $(f \leqslant \alpha) \triangleq \{x \in \mathbb{R}^n \mid f(x) \leqslant \alpha\}$ is convex. Is the converse true? Why?

12. Let $f : \mathbb{R}^n \to \mathbb{R}$. Show that $f(\cdot)$ is both convex and concave if and only there are some $a \in \mathbb{R}^n$ and $b \in \mathbb{R}$ such that

$$f(x) = a^\top x + b.$$

13. Let $f : \mathbb{R}^n \to \mathbb{R}$ be convex with $f(0) = 0$. Show that for any $k \in (0,1)$, and $x \in \mathbb{R}^n$, $f(kx) \leqslant kf(x)$.

2 Optimization Problems under Convexity Conditions

In this section, we consider Problem (G) with the condition that the feasible set F is convex and the objective function $f(\cdot)$ is convex. Our first result is the following.

Theorem 2.1. Let $F \subseteq \mathbb{R}^n$ be a convex set and $f : F \to \mathbb{R}$ be a convex function.

(i) Let $\mathcal{S} \subseteq F$ be the set of all global minimum points of Problem (G). Then $\mathcal{S}$ is convex. Moreover, any local minimum point $\bar{x}$ of Problem (G) must be in $\mathcal{S}$, i.e., any local minimum point must be a global minimum point. Furthermore, if $f(\cdot)$ is strictly convex, then $\mathcal{S}$ consists of at most one point.

(ii) If $f(\cdot)$ is C^1, then any point $\bar{x} \in F$ is a (global) minimum point of Problem (G) if and only if

$$f_x(\bar{x})(y - \bar{x}) \geqslant 0, \qquad \forall y \in F. \tag{2.1}$$

(iii) If F is compact, then $f(\cdot)$ attains its maximum value over F on the boundary ∂F of F.

Proof. (i) If $\mathcal{S} = \varnothing$, we are done. Suppose $\mathcal{S} \neq \varnothing$. Let c_0 be the minimum value of $f(\cdot)$ over F. Then

$$\mathcal{S} = \{x \in F \mid f(x) \leqslant c_0\}.$$

By convexity of $f(\cdot)$, we can easily check that $\mathcal{S}$ is convex.

Now, suppose $\bar{x} \in F$ is a local minimum point, and there exists a $y \in F$ such that

$$f(y) < f(\bar{x}).$$

Then for $\varepsilon \in (0,1)$ small enough, $\bar{x} + \varepsilon(y - \bar{x})$ is in a small neighborhood of $\bar{x}$, and

$$f(\bar{x} + \varepsilon(y - \bar{x})) = f((1 - \varepsilon)\bar{x} + \varepsilon y) \leqslant (1 - \varepsilon)f(\bar{x}) + \varepsilon f(y) < f(\bar{x}),$$

contradicting the local optimality of $\bar{x}$.

Now, suppose $\bar{x}, \bar{y} \in \mathcal{S}$, which are different. Then $\lambda \bar{x} + (1 - \lambda)\bar{y} \in F$ and due to the strict convexity of $f(\cdot)$, we have

$$f(\lambda \bar{x} + (1 - \lambda)\bar{y}) < \lambda f(\bar{x}) + (1 - \lambda)f(\bar{y}) = \inf_{x \in F} f(x),$$

contradicting the optimality of $\bar{x}$ and $\bar{y}$. Thus, $\mathcal{S}$ contains at most one point.

(ii) Suppose $\bar{x} \in F$ is a minimum point of $f(\cdot)$. Then for any $y \in F$,

$$0 \leqslant \lim_{\varepsilon \to 0} \frac{f(\bar{x} + \varepsilon(y - \bar{x})) - f(\bar{x})}{\varepsilon} = f_x(\bar{x})(y - \bar{x}),$$

which gives the necessity of (2.1). Conversely, suppose (2.1) holds. Then for any $y \in F$, by convexity of $f(\cdot)$ (using (1.19)), one has

$$f(y) \geqslant f(\bar{x}) + f_x(\bar{x})(y - \bar{x}) \geqslant f(\bar{x})$$

proving the minimality of $\bar{x}$.

(iii) Suppose $\bar{x}$ is a (global) maximum of $f(\cdot)$ over F. If $\bar{x}$ is an interior point of F, then we take any straight line L passing through $\bar{x}$. Since F is bounded, the line L will intersect the boundary ∂F of F at two different points y_1, y_2. Hence, we must have

$$\bar{x} = \alpha y_1 + (1 - \alpha)y_2,$$

for some $\alpha \in (0, 1)$. Then by the convexity of $f(\cdot)$, one has

$$f(\bar{x}) = f(\alpha y_1 + (1 - \alpha)y_2) \leqslant \alpha f(y_1) + (1 - \alpha)f(y_2) \leqslant f(\bar{x}).$$

This means that both y_1 and y_2 must be maximum points as well. $\qquad\square$

Note that the convexity condition of F is almost necessary for the uniqueness of the minimum. Here is an easy counterexample: Let $f(x) = \|x\|^2$, and $F = \{x \in \mathbb{R}^n \mid \|x\| = 1\}$ which is not convex. All the points on F are optimal although $f(\cdot)$ is strictly convex.

Also, the strict convexity of $f(\cdot)$ is almost necessary for the uniqueness of the minimum. For example, if $f(x, y) = x$ which is not strictly convex, and $F = [-1, 1] \times [-1, 1]$, then every point $(-1, y) \in \{-1\} \times [-1, 1]$ is optimal.

We now look at the following NLP problem:

$$\begin{cases} \min \quad f(x), \\ \text{subject to} \quad Ax = b, \quad h(x) \leqslant 0. \end{cases} \tag{2.2}$$

Here, we assume the following:

$$\begin{cases} G \subseteq \mathbb{R}^n \text{ is convex and closed}, \\ f, h^j : G \to \mathbb{R} \text{ are convex and } C^1, \quad 1 \leqslant j \leqslant \ell, \\ A \in \mathbb{R}^{m \times n}, \quad b \in \mathbb{R}^m. \end{cases} \tag{2.3}$$

It is not hard to see that under the above conditions, the feasible set F is convex and closed, the objective function is convex. Therefore, we refer to (2.2) as a *convex programming* problem.

Definition 2.2. (i) Problem (2.2) is said to be *consistent* if the feasible set $F \equiv \{x \in \mathbb{R}^n \mid Ax = b, \ h(x) \leqslant 0\}$ is non-empty.

(ii) Problem (2.2) is said to be *strongly consistent* (or to satisfy the *Slater condition*) if there exists an $x_0 \in \mathbb{R}^n$ such that $Ax_0 = b$ and $h(x_0) < 0$.

Note that due to the equality constraint, in order the feasible set to be non-trivial (meaning that it contains more than one point), and there are no redundant equality constraints, we need to assume, hereafter, that

$$m = \operatorname{rank}(A, b) = \operatorname{rank}(A) < n. \tag{2.4}$$

We have the following result.

Proposition 2.3. *Let $G \subseteq \mathbb{R}^n$ be convex, $f : G \to \mathbb{R}$ and $h : G \to \mathbb{R}^\ell$ be convex, $A \in \mathbb{R}^{m \times n}$ with $\operatorname{rank}(A) = m < n$, and $b \in \mathbb{R}^m$. Let $\bar{x}$ be feasible. Then the MF qualification condition holds at $\bar{x}$ if and only if Problem (2.2) is strongly consistent.*

Proof. $\Rightarrow$: Suppose for Problem (2.2), the MF qualification condition holds at the feasible point $\bar{x}$. By Definition 4.5 of Chapter 3, with $g(x) = Ax - b$, we see that $A = g_x(\bar{x})$ is of full rank (which has been assumed) and there exists a $z \in \mathbb{R}^n$ such that $Az = 0$ and

$$h_x^J(\bar{x})z < 0,$$

with

$$\begin{cases} J(\bar{x}) = \{j \mid h^j(\bar{x}) = 0, \ 1 \leqslant j \leqslant \ell\} \equiv \{j_1, \cdots, j_k\}, \\ h^J(x) = (h^{j_1}(x), \cdots, h^{j_k}(x))^\top. \end{cases}$$

Without loss of generality, we may let

$$J = \{1, 2, \cdots, k\}, \qquad J^c = \{k+1, \cdots, \ell\}.$$

Thus,

$$h^J(x) = (h^1(x), \cdots, h^k(x))^\top, \qquad h^{J^c}(x) = (h^{k+1}(x), \cdots, h^\ell(x))^\top,$$
$$h^J(\bar{x}) = 0, \qquad h^{J^c}(\bar{x}) < 0.$$

Then, for small enough $\delta > 0$, we have

$$h(\bar{x} + \delta z) = \begin{pmatrix} \delta h_x^J(\bar{x})z \\ h^{J^c}(\bar{x}) + \delta h^{J^c}(\bar{x})z \end{pmatrix} + o(\delta) < 0,$$

and

$$A(\bar{x} + \delta z) = b.$$

Hence, by choosing $x_0 = \bar{x} + \delta z$, we see that Problem (2.2) is strongly consistent.

$\Leftarrow$: Suppose Problem (2.2) is stronger consistent, i.e., there exists an $x_0 \in \mathbb{R}^n$ such that

$$A x_0 = b, \qquad h(x_0) < 0.$$

Let $\bar{x}$ be feasible. If $h(\bar{x}) < 0$, then $J(\bar{x}) = \varnothing$, by rank $(A) = m$, the constraint is regular at $\bar{x}$, which implies that the MF qualification condition holds at $\bar{x}$ (by Proposition 4.8 of Chapter 3). We now, let $J(\bar{x}) \neq \varnothing$. Then let $z = x_0 - \bar{x}$. We have

$$Az = A x_0 - A \bar{x} = 0,$$

and for each $i \in J(\bar{x})$,

$$0 > h^i(x_0) = h^i(\bar{x} + \delta z) = h^i(\bar{x}) + \delta h_x^i(\bar{x})z + o(\delta) = \delta h_x^i(\bar{x})z + o(\delta).$$

Hence, it is necessary that

$$h_x^J(\bar{x})z < 0.$$

This means that the MF qualification condition holds at $\bar{x}$. $\qquad\square$

We now look at necessary conditions for optimal solutions to Problem (2.2). Note that in the current case the Fritz John's function is given by the following:

$$\Lambda(x, \lambda^0, \lambda, \mu) = \lambda^0 f(x) + \lambda^\top(Ax - b) + \mu^\top h(x),$$
$$\forall (x, \lambda^0, \lambda, \mu) \in G \times [0, 1] \times \mathbb{R}^m \times \mathbb{R}_+^\ell. \tag{2.5}$$

Our first result is a kind of restatement of Fritz John's necessary condition (see Theorem 4.3 of Chapter 3).

Theorem 2.4. *Let* (2.3) *hold and* $\bar{x}$ *be an optimal solution to Problem* (2.2). *Then there exists a non-zero triple* $(\bar{\lambda}^0, \bar{\lambda}, \bar{\mu}) \in [0,1] \times \mathbb{R}^m \times \mathbb{R}^\ell_+$ *such that*

$$\bar{\mu}^\top h(\bar{x}) = 0, \tag{2.6}$$

and

$$\Lambda(\bar{x}, \bar{\lambda}^0, \lambda, \mu) \leqslant \Lambda(\bar{x}, \bar{\lambda}^0, \bar{\lambda}, \bar{\mu}) \leqslant \Lambda(x, \bar{\lambda}^0, \bar{\lambda}, \bar{\mu}), \tag{2.7}$$
$$\forall (x, \lambda, \mu) \in G \times \mathbb{R}^m \times \mathbb{R}^\ell_+.$$

Moreover,

$$\Lambda(\bar{x}, \bar{\lambda}^0, \bar{\lambda}, \bar{\mu}) = \bar{\lambda}^0 f(\bar{x}). \tag{2.8}$$

Further, if Problem (2.2) *is strongly consistent,* (2.4) *holds, and* $G = \mathbb{R}^n$, *then* $\bar{\lambda}^0 > 0$.

Proof. By Fritz John's Theorem (Theorem 4.3 of Chapter 3), we know that there exists a non-zero triple $(\bar{\lambda}^0, \bar{\lambda}, \bar{\mu}) \in [0,1] \times \mathbb{R}^m \times \mathbb{R}^\ell_+$ such that

$$0 = \bar{\lambda}^0 f_x(\bar{x}) + \bar{\lambda}^\top A + \bar{\mu}^\top h_x(\bar{x}), \tag{2.9}$$

and (2.6) holds. Now, since $f(\cdot)$ and all $h^j(\cdot)$ are convex, we have

$$\begin{cases} f(x) \geqslant f(\bar{x}) + f_x(\bar{x})(x - \bar{x}), \\ h^j(x) \geqslant h^j(\bar{x}) + h_x^j(\bar{x})(x - \bar{x}), & 1 \leqslant j \leqslant \ell. \end{cases}$$

Therefore, by $\bar{\lambda}^0 \geqslant 0$ and $\bar{\mu} \geqslant 0$, we have (note $A\bar{x} = b$)

$$\Lambda(x, \bar{\lambda}^0, \bar{\lambda}, \bar{\mu}) = \bar{\lambda}^0 f(x) + \bar{\lambda}^\top (Ax - b) + \bar{\mu}^\top h(x)$$
$$\geqslant \bar{\lambda}^0 [f(\bar{x}) + f_x(\bar{x})(x - \bar{x})] + \bar{\lambda}^\top [A(x - \bar{x})] + \bar{\mu}^\top [h(\bar{x}) + h_x(\bar{x})(x - \bar{x})]$$
$$= \bar{\lambda}^0 f(\bar{x}) + [\bar{\lambda}^0 f_x(\bar{x}) + \bar{\lambda}^\top A + \bar{\mu}^\top h_x(\bar{x})](x - \bar{x}) + \bar{\mu}^\top h(\bar{x}) = \bar{\lambda}^0 f(\bar{x})$$
$$= \bar{\lambda}^0 f(\bar{x}) + \bar{\lambda}^\top (A\bar{x} - b) + \bar{\mu}^\top h(\bar{x}) = \Lambda(\bar{x}, \bar{\lambda}^0, \bar{\lambda}, \bar{\mu})$$
$$\geqslant \bar{\lambda}^0 f(\bar{x}) + \lambda^\top (A\bar{x} - b) + \mu^\top h(\bar{x}) = \Lambda(\bar{x}, \bar{\lambda}^0, \lambda, \mu).$$

Here, we note that $h(\bar{x}) \leqslant 0$ and $\mu \geqslant 0$. The above proves (2.7) and (2.8).

Now, suppose Problem (2.2) is strongly consistent, but $\bar{\lambda}^0 = 0$. Then

$$\bar{\lambda}^\top (Ax - b) + \bar{\mu}^\top h(x) \geqslant 0, \qquad \forall x \in G. \tag{2.10}$$

We claim that $\bar{\mu} \neq 0$. Otherwise, the above leads to

$$(A^\top \bar{\lambda})^\top x = \bar{\lambda}^\top Ax \geqslant \bar{\lambda}^\top b, \qquad \forall x \in \mathbb{R}^n.$$

Thus, we must have $A^\top \bar\lambda = 0$. Then $AA^\top \bar\lambda = 0$, this leads to $\bar\lambda = 0$ since $\operatorname{rank}(A) = m$ which implies the invertibility of $AA^\top \in \mathbb{R}^{m\times m}$. Consequently, $(\bar\lambda^0, \bar\lambda, \bar\mu) = 0$, a contradiction. Hence, $\bar\mu \neq 0$.

Now by the strong consistency of the Problem (2.2), we have some $x_0 \in \mathbb{R}^n$ such that $Ax_0 = b$ and $h(x_0) < 0$. Thus, taking $x = x_0$ in (2.10), we have

$$0 \leqslant \bar\mu^T h(x_0) < 0,$$

a contradiction. Hence, $\bar\lambda^0 > 0$ must be true. □

The above means that if $\bar x$ is an optimal solution, and if $(\bar\lambda^0, \bar\lambda, \bar\mu)$ is the corresponding Frtiz John's multiplier, then $(\bar x, \bar\lambda, \bar\mu)$ is a saddle point of the map $(x, \lambda, \mu) \mapsto \Lambda(x, \bar\lambda^0, \lambda, \mu)$ over $G \times \mathbb{R}^m \times \mathbb{R}_+^\ell$. Note that in the general case discussed in the Section 4 of Chapter 3, we did not have the saddle point of the Fritz John's function as a necessary condition of optimal solution of NLP problems. Further, from the above, we see that in the case that the problem is strictly consistent and $G = \mathbb{R}^n$, one has $\bar\lambda^0 > 0$. Thus, by scaling, we may let $\bar\lambda^0 = 1$. Then the above result says that there exists a pair $(\bar\lambda, \bar\mu) \in \mathbb{R}^m \times \mathbb{R}_+^\ell$ such that (2.6) holds and

$$L(\bar x, \lambda, \mu) \leqslant L(\bar x, \bar\lambda, \bar\mu) \leqslant L(x, \bar\lambda, \bar\mu), \quad \forall (x, \lambda, \mu) \in \mathbb{R}^n \times \mathbb{R}^m \times \mathbb{R}_+^\ell, \quad (2.11)$$

and

$$f(\bar x) = L(\bar x, \bar\lambda, \bar\mu). \tag{2.12}$$

The following gives a sufficient condition for optimal solution to Problem (2.2).

Theorem 2.5. *Let* (2.3) *hold. Suppose* $(\bar x, \lambda, \mu) \in G \times \mathbb{R}^m \times \mathbb{R}_+^\ell$ *satisfies the following:*

$$\begin{cases} A\bar x = b, & h(\bar x) \leqslant 0, \\ \mu^\top h(\bar x) = 0, \\ f_x(\bar x) + \lambda^\top A + \mu^\top h_x(\bar x) = 0. \end{cases} \tag{2.13}$$

Then $\bar x \in G$ *is a solution of Problem* (2.2).

Proof. By the convexity of $h(\cdot)$, we have

$$h(x) - h(\bar x) \geqslant h_x(\bar x)(x - \bar x), \quad \forall x \in G.$$

Thus, it follows from $\mu \geqslant 0$ that

$$\mu^\top h_x(\bar x)(x - \bar x) \leqslant \mu^\top [h(x) - h(\bar x)], \quad x \in G.$$

Hence, by the convexity of $f(\cdot)$, one sees that for any $x \in G$, with $g(x) = Ax - b = 0$ and $h(x) \leqslant 0$,

$$f(x) - f(\bar{x}) \geqslant f_x(\bar{x})(x - \bar{x}) = -[\lambda^\top A + \mu^\top h_x(\bar{x})](x - \bar{x})$$
$$\geqslant -\lambda^\top(Ax - A\bar{x}) - \mu^\top[h(x) - h(\bar{x})] \geqslant 0,$$

which means that $\bar{x}$ is a global minimum of Problem (2.2). $\square$

Exercises

1. Let $Q \in \mathbb{S}^n$ be positive definite and let $b \in \mathbb{R}^n$. Consider the following optimization problem:

$$\begin{cases} \min \quad x^\top Q x, \\ \text{subject to } b^\top x \leqslant 1. \end{cases}$$

Show that the above problem admits a unique optimal solution.

2. Let $Q \in \mathbb{S}^n$ be positive definite and let $c \in \mathbb{R}^n$. Consider the following optimization problem:

$$\begin{cases} \min \quad c^\top x, \\ \text{subject to } x^\top Q x \leqslant 1. \end{cases}$$

Show that the above problem admits an optimal solution.

3. Let $F_1, F_2 \subseteq \mathbb{R}^n$ and $F_1 \subseteq F_2^\circ$. Let $f : F_2 \to \mathbb{R}$. Consider optimization problems corresponding to $(F_1, f(\cdot))$ and $(F_2, f(\cdot))$. Suppose both problem admit optimal solutions. What can you say about these solutions? Give your reason. What happens if F_1 and F_2 are convex sets and $f(\cdot)$ is a convex function?

4. Solving the following convex programming problems:

(i) Minimize $x^2 + y^2 + 2x + 2y$, subject to $x + y = 0$, $x \leqslant 0$.

(ii) Minimize $(x - 1)^2 + y^2 + x$, subject to $x - y = 1$, $x^2 + y^2 - 1 \leqslant 0$.

5. Let $f : \mathbb{R}^n \to \mathbb{R}$ be continuous and $h : \mathbb{R}^n \to \mathbb{R}$ be C^2 such that for some $\delta > 0$,

$$h_{xx}(x) \geqslant \delta I, \qquad \forall x \in \mathbb{R}^n.$$

Suppose $F = \{x \in \mathbb{R}^n \mid h(x) \leqslant 0\}$ is non-empty. Show that $f(\cdot)$ admits a minimum over F.

3 Lagrange Duality

Let $G = \mathbb{R}^n$ and $f : \mathbb{R}^n \to \mathbb{R}$, $A \in \mathbb{R}^{m \times n}$, $b \in \mathbb{R}^m$, $h : \mathbb{R}^n \to \mathbb{R}^\ell$, with $f(\cdot)$ and $h(\cdot)$ convex. Denote

$$F = \{x \in \mathbb{R}^n \mid Ax = b, \ h(x) \leqslant 0\}. \tag{3.1}$$

Define

$$
\begin{aligned}
\theta(\lambda, \mu) &= \inf_{x \in \mathbb{R}^n} \left[f(x) + \lambda^\top (Ax - b) + \mu^\top h(x) \right] \\
&\equiv \inf_{x \in \mathbb{R}^n} L(x, \lambda, \mu), \quad \forall (\lambda, \mu) \in \mathbb{R}^m \times \mathbb{R}^\ell_+,
\end{aligned}
\tag{3.2}
$$

and

$$\Phi = \{(\lambda, \mu) \in \mathbb{R}^m \times \mathbb{R}^\ell_+ \mid \theta(\lambda, \mu) > -\infty\}. \tag{3.3}$$

In the case that $\Phi \neq \varnothing$, we may pose the following problem:

$$
\begin{cases}
\max \quad \theta(\lambda, \mu), \\
\text{subject to} \quad (\lambda, \mu) \in \Phi.
\end{cases}
\tag{3.4}
$$

This problem is called the *dual problem* of Problem (2.2) which is often referred to as the *primal problem*. Note that the formulation of Problem (3.4) does not need the strong consistency or the existence of an optimal solution of the primal problem. The following result tells us how the dual problem gives information about the primal problem and vice-versa. Such a result is called the *Lagrange duality*.

Theorem 3.1. (i) *If $F \neq \varnothing$ and $\Phi \neq \varnothing$, then*

$$\theta(\lambda, \mu) \leqslant f(x), \qquad \forall x \in F, \ (\lambda, \mu) \in \Phi. \tag{3.5}$$

Consequently, if we define

$$\bar{\theta} = \sup_{(\lambda, \mu) \in \Phi} \theta(\lambda, \mu), \qquad \bar{f} = \inf_{x \in F} f(x), \tag{3.6}$$

then both $\bar{\theta}$ and $\bar{f}$ are finite and

$$\bar{\theta} \leqslant \bar{f}. \tag{3.7}$$

(ii) *Let $\Phi \neq \varnothing$. If $\theta(\lambda, \mu)$ is unbounded above, then $F = \varnothing$.*

(iii) *Let $F \neq \varnothing$. If $f(x)$ is unbounded below, then $\Phi = \varnothing$.*

(iv) *If there exists a $(\bar{\lambda}, \bar{\mu}) \in \Phi$ and an $\bar{x} \in F$ such that*

$$f(\bar{x}) = \theta(\bar{\lambda}, \bar{\mu}), \tag{3.8}$$

then

$$\bar{\theta} = \theta(\bar{\lambda}, \bar{\mu}) = f(\bar{x}) = \bar{f}. \tag{3.9}$$

In this case, $\bar{x}$ is a solution of Problem (2.2) and $(\bar{\lambda}, \bar{\mu})$ is a solution of Problem (3.4).

Proof. (i) We note that for any $x \in F$ and $(\lambda, \mu) \in \Phi$, we have

$$Ax - b = 0, \quad h(x) \leqslant 0, \quad \mu \geqslant 0.$$

Hence,

$$f(x) + \lambda^\top(Ax - b) + \mu^\top h(x) \leqslant f(x). \tag{3.10}$$

Taking infimum over $x \in \mathbb{R}^n$ on the left hand side, we obtain inequality (3.5). Then the finiteness of $\bar\theta$ and $\bar f$ is clear and (3.7) follows easily.

(ii) If $F \neq \varnothing$, then $\theta(\lambda, \mu)$ would be bounded above by (i). Hence, if $\theta(\lambda, \mu)$ is unbounded above, one has to have $F = \varnothing$.

(iii) If $\Phi \neq \varnothing$, then $f(x)$ would be bounded below by (i). Hence, if $f(x)$ is unbounded below, one has to have $\Phi = \varnothing$.

(iv) Now, if $\bar x \in F$ and $(\bar\lambda, \bar\mu) \in \Phi$ such that (3.8) holds, then from the chain of inequalities

$$\theta(\bar\lambda, \bar\mu) \leqslant \bar\theta \leqslant \bar f \leqslant f(\bar x) = \theta(\bar\lambda, \bar\mu),$$

we obtain (3.9). □

The above result naturally leads to the following notion.

Definition 3.2. The number $\bar f - \bar\theta \geqslant 0$ is called the *duality gap* of the primal problem (2.2) and its dual problem (3.4).

We have the following result.

Theorem 3.3. *Let the primal problem (2.2) be strongly consistent and admit a solution. Then*

(i) *The duality gap is zero.*

(ii) $(\bar\lambda, \bar\mu)$ *is a KKT multiplier of the primal problem (2.2) if and only if* $(\bar\lambda, \bar\mu)$ *is a solution of the dual problem (3.4).*

Proof. (i) Let $\bar x \in F$ be a solution to the primal problem (2.2). By the strong consistency of the primal problem, we may take $\lambda^0 = 1$ in the Fritz John's multiplier, which means that for some KKT multiplier $(\bar\lambda, \bar\mu)$, one has (making use of Theorem 2.4)

$$\begin{aligned} f(\bar x) &= f(\bar x) + \bar\lambda^\top(A\bar x - b) + \bar\mu^\top h(\bar x) = L(\bar x, \bar\lambda, \bar\mu) \\ &\leqslant L(x, \bar\lambda, \bar\mu) = f(x) + \bar\lambda^\top(Ax - b) + \bar\mu^\top h(x), \qquad \forall x \in \mathbb{R}^n. \end{aligned}$$

Hence, by the definitions of $\theta(\cdot,\cdot)$, $\bar{\theta}$ and $\bar{f}$, together with Theorems 3.1 and 2.4, we have

$$\bar{\theta} \leqslant \bar{f} = \inf_{x \in F} f(x) = f(\bar{x}) \leqslant L(x, \bar{\lambda}, \bar{\mu}), \qquad \forall x \in \mathbb{R}^n.$$

Therefore,

$$\bar{\theta} \leqslant \bar{f} \leqslant \inf_{x \in \mathbb{R}^n} L(x, \bar{\lambda}, \bar{\mu}) = \theta(\bar{\lambda}, \bar{\mu}) \leqslant \sup_{(\lambda, \mu) \in \mathbb{R}^m \times \mathbb{R}^\ell_+} \theta(\lambda, \mu) = \bar{\theta}.$$

Consequently, all the equalities hold in the above. In particular, we have $\bar{\theta} = \bar{f}$, the duality gap is zero.

(ii) Suppose $(\bar{\lambda}, \bar{\mu})$ is a KKT multiplier of the primal problem (2.2). Then from the proof of part (i), we see that $\bar{f} = f(\bar{x}) = \theta(\bar{\lambda}, \bar{\mu}) = \bar{\theta}$. This implies

$$\theta(\bar{\lambda}, \bar{\mu}) \leqslant \sup_{(\lambda, \mu) \in \mathbb{R}^m \times \mathbb{R}^\ell_+} \theta(\lambda, \mu) \leqslant \bar{f} = \theta(\bar{\lambda}, \bar{\mu}).$$

Therefore, $(\bar{\lambda}, \bar{\mu})$ is a solution of the dual problem.

Conversely, if $(\bar{\lambda}, \bar{\mu}) \in \mathbb{R}^m \times \mathbb{R}^\ell_+$ is a solution to the dual problem, then we have

$$\inf_{x \in \mathbb{R}^n} \left[f(x) + \bar{\lambda}^\top (Ax - b) + \bar{\mu}^\top h(x) \right] = \theta(\bar{\lambda}, \bar{\mu}) = \bar{\theta} = \bar{f} = f(\bar{x}).$$

This means that $x \mapsto f(x) + \bar{\lambda}^\top (Ax - b) + \bar{\mu}^\top h(x)$ attains its minimum at $x = \bar{x}$. Hence,

$$f_x(\bar{x}) + \bar{\lambda}^\top A + \bar{\mu}^\top h_x(\bar{x}) = 0,$$

and

$$f(\bar{x}) + \bar{\lambda}^\top (A\bar{x} - b) + \bar{\mu}^\top h(\bar{x}) = L(\bar{x}, \bar{\lambda}, \bar{\mu})$$
$$\geqslant \inf_{x \in \mathbb{R}^n} L(x, \bar{\lambda}, \bar{\mu}) = \theta(\bar{\lambda}, \bar{\mu}) = f(\bar{x}),$$

which implies (noting $\bar{x} \in F$)

$$\bar{\mu}^\top h(\bar{x}) \geqslant 0.$$

But, $h(\bar{x}) \leqslant 0$ and $\bar{\mu} \geqslant 0$. Hence, we must have

$$\bar{\mu}^\top h(\bar{x}) = 0.$$

This means that $(\bar{\lambda}, \bar{\mu})$ is a KKT multiplier. $\qquad\square$

Note that the above Lagrange duality results hold for the case $m = 0$ or $\ell = 0$. In other words, the results remain true if either the equality constraint or the inequality constraint is absent.

We now present some examples. The first example shows that the dual gap could be positive, even if the primal problem admits an optimal solution.

Example 3.4. Consider the following problem.

$$\begin{cases} \min \quad f(x,y) = e^{-y}, \\ \text{subject to} \quad h(x,y) = \sqrt{x^2 + y^2} - x \leqslant 0. \end{cases}$$

Then it is easy to see that the feasible set is

$$F = \{(x,y) \in \mathbb{R}^2 \mid x \geqslant 0, \ y = 0\}.$$

Hence, any point in F is an optimal solution to the problem. Now,

$$L(x,y,\mu) = e^{-y} + \mu\left(\sqrt{x^2 + y^2} - x\right)$$

$$= e^{-y} + \mu\frac{y^2}{\sqrt{x^2 + y^2} + x} \geqslant 0,$$

$$\forall(x,y) \in \mathbb{R}^2, \ \mu \geqslant 0.$$

By taking $x = y^3$ and letting $y \to \infty$, we see that $L(y^3, y, \mu) \to 0$, Therefore,

$$\theta(\mu) = \inf_{(x,y)\in\mathbb{R}^2} \left[e^{-y} + \mu\left(\sqrt{x^2 + y^2} - x\right)\right] = 0, \quad \forall \mu \geqslant 0.$$

But, $\bar{f} = f(0,0) = 1 > 0 = \bar{\theta}$. Then, the dual gap $\bar{f} - \bar{\theta} = 1$ is positive, and both the primal problem and dual problem admit solutions. Note that in this example, the primal problem is not strongly consistent.

The following example shows that the solvability of the primal problem and the dual gap being zero does not necessarily imply the solvability of the dual problem (which is guaranteed when the primal problem is strongly consistent).

Example 3.5. Consider the following primal problem:

$$\begin{cases} \min \quad f(x,y) = x, \\ \text{subject to} \quad h(x,y) \equiv \begin{pmatrix} y \\ x^2 - y \end{pmatrix} \leqslant 0. \end{cases}$$

Then the feasible set $F = \{(0,0)\}$. Therefore,

$$\bar{f} = \min_{(x,y)\in F} f(x,y) = f(0,0) = 0.$$

Now, the KKT function is given by

$$L(x,y,\mu_1,\mu_2) = x + \mu_1 y + \mu_2(x^2 - y) = x + \mu_2 x^2 + (\mu_1 - \mu_2)y.$$

Hence,

$$\Phi = \{(\mu_1, \mu_2) \geq 0 \mid \mu_1 = \mu_2 > 0\},$$

and for any $\mu_1 = \mu_2 > 0$,

$$\theta(\mu_1, \mu_2) = \inf_{x \in \mathbb{R}} (x + \mu_2 x^2) = -\frac{1}{4\mu_2}, \qquad \mu_2 > 0.$$

Thus,

$$\bar{\theta} = \sup_{(\mu_1, \mu_2) \in \Phi} \theta(\mu_1, \mu_2) = 0,$$

which is not attained. This means that the dual problem does not have a solution, although the dual gap is zero. Again, we note that the primal problem is not strongly consistent.

Next, we know that when the primal problem is solvable, the strong consistency of the primal problem is sufficient for the dual gap to be zero and the solvability of dual problem. The following shows that the strong consistency is not necessary for these.

Example 3.6. Consider the following primal problem:

$$\begin{cases} \min \quad f(x, y) = x, \\ \text{subject to} \quad h(x, y) = \begin{pmatrix} (x+1)^2 + y^2 - 1 \\ -x \end{pmatrix} \leqslant 0. \end{cases}$$

Then the feasible set $F = \{(0, 0)\}$, and

$$\bar{f} = f(0, 0) = 0.$$

Let

$$L(x, y, \mu_1, \mu_2) = x + \mu_1 \left((x+1)^2 + y^2 - 1 \right) - \mu_2 x,$$

$$\mu_1, \mu_2 \geqslant 0, \ (x, y) \in \mathbb{R}^2,$$

and

$$\theta(\mu_1, \mu_2) = \inf_{(x,y) \in \mathbb{R}^2} \left[x + \mu_1 \left((x+1)^2 + y^2 - 1 \right) - \mu_2 x \right]$$

$$= \inf_{x \in \mathbb{R}} \left[\mu_1 x^2 + (1 + 2\mu_1 - \mu_2) x \right] = \begin{cases} -\dfrac{(1 + 2\mu_1 - \mu_2)^2}{4\mu_1}, & \text{if } \mu_1 > 0, \\ -\infty, & \text{if } \mu_1 = 0. \end{cases}$$

Note that

$$0 = \bar{f} \geqslant \bar{\theta} = \sup_{(\mu_1, \mu_2) \in \Phi} \theta(\mu_1, \mu_2) = \theta(\mu_1, 1 + 2\mu_1) = 0, \qquad \forall (x, y) \in \mathbb{R}^2.$$

Hence, the dual gap is zero and $(\mu_1, 1 + 2\mu_1)$ $(\mu_1 > 0)$ is the solution to the dual problem. Since F is a singleton, the primal problem is not strongly consistent.

Exercises

1. Let $f : G \to \mathbb{R}$ and $g : G \to \mathbb{R}^m$ be given functions. Let $\theta : \mathbb{R}^m \to \mathbb{R}$ be defined by

$$\theta(\lambda) = \inf_{x \in G}[f(x) + \lambda^\top g(x)], \qquad \lambda \in \mathbb{R}^m.$$

(i) Find conditions for $f(\cdot)$ and $g(\cdot)$ so that $\theta(\cdot)$ is well-defined on $\mathbb{R}^m$.

(ii) Show that $\theta(\cdot)$ is concave.

2. Let $f(x, y) = x^2 + y^2 + 2x$, $g(x, y) = x + 2y$, and $h(x, y) = (x - 1)^2 + y^2 - 1$. Calculate

$$\theta(\lambda, \mu) = \inf_{(x,y) \in \mathbb{R}^2}\Big[f(x, y) + \lambda g(x, y) + \mu h(x, y)\Big], \quad (\lambda, \mu) \in \mathbb{R} \times [0, \infty).$$

Then solve the dual problem for the corresponding Problem (G) associated with $f(\cdot), g(\cdot), h(\cdot)$.

4   Quasi-Convexity and Related Optimization Problems

We now would like to extend the notion of convex function.

Definition 4.1. Let $G \subseteq \mathbb{R}^n$ be convex.

(i) A function $f : G \to \mathbb{R}$ is said to be *quasi-convex* on G if

$$f(\lambda x + (1 - \lambda)y) \leqslant \max\{f(x), f(y)\}, \qquad x, y \in G, \ \lambda \in (0, 1). \qquad (4.1)$$

If the strict inequality holds in the above, we call $f(\cdot)$ a *strictly quasi-convex* function.

(ii) A function $f : G \to \mathbb{R}$ is said to be *quasi-concave* on G if

$$f(\lambda x + (1 - \lambda)y) \geqslant \min\{f(x), f(y)\}, \qquad x, y \in G, \ \lambda \in (0, 1). \qquad (4.2)$$

If the strict inequality holds in the above, we call $f(\cdot)$ a *strict quasi-concave* function.

Let us make some observations. First of all, $f(\cdot)$ is quasi-convex on a convex set G if and only if $-f(\cdot)$ is quasi-concave on G. Secondly, note that for $f : G \to \mathbb{R}$ with $G \subseteq \mathbb{R}^n$ being convex, we have

$$\min\{f(x), f(y)\} \leqslant \lambda f(x) + (1 - \lambda)f(y) \leqslant \max\{f(x), f(y)\}, \qquad (4.3)$$
$$\forall x, y \in G, \ \lambda \in (0, 1).$$

This implies that any convex function is quasi-convex, and any concave function is quasi-concave. Also, the above tells us that for the case

$n = 1$, any monotone function (either non-decreasing or non-increasing) $f : [a, b] \to \mathbb{R}$ is both quasi-convex and quasi-concave. In particular, any monotone piecewise constant functions are both quasi-convex and quasi-concave. Hence, we at least have the following features for the quasi-convex functions:

- Quasi-convex function could be discontinuous;

- Quasi-convex function could have local minima that are not global minima;

- Quasi-convex function could be not convex.

Quasi-concave functions have similar features.

Now, we present a characterization of quasi-convex and quasi-concave functions.

Proposition 4.2. *Let $G \subseteq \mathbb{R}^n$ be convex.*

(i) *Function $f : G \to \mathbb{R}$ is quasi-convex if and only if for any $\alpha \in \mathbb{R}$, the lower-contour set*

$$(f \leqslant \alpha) \equiv \{x \in G \mid f(x) \leqslant \alpha\} \tag{4.4}$$

is convex.

(ii) *Function $f : G \to \mathbb{R}$ is quasi-concave if and only if for any $\alpha \in \mathbb{R}$, the upper-contour set*

$$(f \geqslant \alpha) \equiv \{x \in G \mid f(x) \geqslant \alpha\} \tag{4.5}$$

is convex.

Proof. We only prove (i). Suppose $f(\cdot)$ is quasi-convex. For any $\alpha \in \mathbb{R}$, if $(f \leqslant \alpha)$ is empty or a singleton, the conclusion is clear. Now, let $x, y \in (f \leqslant \alpha)$, and $\lambda \in (0, 1)$. Then

$$f(\lambda x + (1 - \lambda)y) \leqslant \max\{f(x), f(y)\} \leqslant \alpha,$$

which means that $(f \leqslant \alpha)$ is convex for any $\alpha \in \mathbb{R}$, proving the necessity.

Conversely, suppose (4.4) holds. Then for any $x, y \in G$ and $\lambda \in (0, 1)$, without loss of generality, we let $f(x) \leqslant f(y) \equiv \alpha$. Thus by the convexity of $(f \leqslant \alpha)$ we have

$$f(\lambda x + (1 - \lambda)y) \leqslant \alpha = f(y) = \max\{f(x), f(y)\},$$

proving the sufficiency. $\qquad\qquad\qquad\qquad\qquad\qquad\qquad\qquad\qquad\square$

We know that when $f : G \to \mathbb{R}$ is convex, the set $(f \leqslant \alpha)$ is convex for all $\alpha \in \mathbb{R}$, and the converse is not necessarily true. Whereas, the above result tells us that the quasi-convexity is equivalent to the convexity of $(f \leqslant \alpha)$ for all $\alpha \in \mathbb{R}$. This enables us to relax conditions of $f(\cdot)$ from convexity to quasi-convexity if we only need the convexity of $(f \leqslant \alpha)$, instead of the convexity of $f(\cdot)$ itself. The same comments apply to the concavity and quasi-concavity.

Next, we note that if $f : G \to \mathbb{R}$ is convex, and $g : \mathbb{R} \to \mathbb{R}$ is non-decreasing and convex, then the composition $g \circ f$ is convex. However, if $g(\cdot)$ is merely non-decreasing, not necessarily convex, then $g \circ f$ is not necessarily convex. It is interesting that we have the following result for quasi-convex functions.

Proposition 4.3. *Let* $f : G \to \mathbb{R}$ *be quasi-convex and* $g : \mathbb{R} \to \mathbb{R}$ *be non-decreasing. Then* $g \circ f : G \to \mathbb{R}$ *is also quasi-convex. In particular, if* $f : G \to \mathbb{R}$ *is convex and* $g : G \to \mathbb{R}$ *is non-decreasing, then* $g \circ f$ *quasi-convex.*

Proof. For any $x, y \in G$ and $\lambda \in (0, 1)$, by the quasi-convexity,

$$f(\lambda x + (1 - \lambda)y) \leqslant \max\{f(x), f(y)\}.$$

Hence, by the monotonicity of $g(\cdot)$, we have

$$g(f(\lambda x + (1 - \lambda)y)) \leqslant g(\max\{f(x), f(y)\}) = \max\{g(f(x)), g(f(y))\},$$

which gives the quasi-convexity of $g \circ f$. ∎

A natural question is whether any quasi-convex function is the composition of a strictly increasing function with a convex function. If this were the case, then any optimization problem with a quasi-convex objective function could be reduced to a problem with convex objective function. Unfortunately, this is not the case, which makes the notion of quasi-convex function alive. We now present an example of quasi-convex function which is not the composition of any non-decreasing function with any convex function.

Example 4.4. Let $f : \mathbb{R}_+ \to \mathbb{R}$ be defined by the following:

$$f(x) = \begin{cases} 0, & x \in [0, 1], \\ -(x - 1)^2, & x \in (1, \infty). \end{cases}$$

Then $f(\cdot)$ is non-increasing, and therefore it is quasi-convex. Suppose there exist a convex function $g(\cdot)$ and a strictly increasing function $\varphi(\cdot)$ such that $f(\cdot) = (\varphi \circ g)(\cdot)$. We first claim that $g(\cdot)$ must be a constant on $[0,1]$. Otherwise, let $x, y \in [0,1]$ with $g(x) > g(y)$. Then

$$0 = f(x) = \varphi(g(x)) > \varphi(g(y)) = f(y) = 0,$$

a contradiction. Next, we claim that $g(\cdot)$ is strictly decreasing on $(1, \infty)$. Otherwise, let us assume $x > y > 1$ such that $g(x) \geqslant g(y)$. Then

$$-(x-1)^2 = f(x) = \varphi(g(x)) \geqslant \varphi(g(y)) = f(y) = -(y-1)^2,$$

which is a contradiction. Hence, $g(\cdot)$ is a convex function such that it is a constant on $[0,1]$ and strictly decreasing on $(1, \infty)$. Then any point $\bar{x} \in (0,1)$ is a local minimum of $g(\cdot)$, which should be a global minimum of $g(\cdot)$ on $[0, \infty)$. But

$$g(\bar{x}) > g(x), \qquad \forall x > 1.$$

This is a contradiction. Therefore, $f(\cdot)$ cannot be written as $(\varphi \circ g)(\cdot)$ for some convex function $g(\cdot)$ and some strictly increasing function $\varphi(\cdot)$.

Now, we present another characterization of quasi-convex functions.

Theorem 4.5. *Let $G \subseteq \mathbb{R}^n$ be an open convex set and $f : G \to \mathbb{R}$ be a C^1 function. Then $f(\cdot)$ is quasi-convex if and only if*

$$f_x(x)(y-x) \leqslant 0, \qquad \forall x, y \in G, \text{ with } f(y) \leqslant f(x). \tag{4.6}$$

Proof. First, suppose $f(\cdot)$ is quasi-convex. Then for any $x, y \in G$ with $f(y) \leqslant f(x)$, and $t \in (0,1)$, we have

$$f(x + t(y-x)) = f((1-t)x + ty) \leqslant \max\{f(x), f(y)\} = f(x).$$

Hence,

$$0 \geqslant \frac{f(x + t(y-x)) - f(x)}{t} \to f_x(x)(y-x).$$

This proves the necessity.

Conversely, suppose (4.6) holds. For any given $x, y \in G$, with $f(y) \leqslant f(x)$, let

$$g(t) = f(x + t(y-x)), \qquad t \in [0,1].$$

Then $g(\cdot)$ is C^1 on $[0,1]$. We claim that

$$g(t) \leqslant g(0), \qquad \forall t \in (0,1), \tag{4.7}$$

which leads to

$$f(x + t(y-x)) = g(t) \leqslant g(0) = f(x) = \max\{f(x), f(y)\},$$

giving the quasi-convexity of $f(\cdot)$. We now prove (4.7) by contradiction. Suppose (4.7) is not true. Then there exists a $t_1 \in (0,1)$ such that $g(t_1) > g(0)$. Let

$$t_0 = \sup\{t \in [0,t_1] \mid g(t) \leqslant g(0)\}.$$

One has that

$$g(t_0) = g(0), \qquad g(t) > g(0), \quad t \in (t_0,t_1].$$

By the Mean Value Theorem, we have some $\bar{t} \in (t_0,t_1)$ such that

$$0 < \frac{g(t_1) - g(t_0)}{t_1 - t_0} = g'(\bar{t}).$$

Since $\bar{t} \in (t_0,t_1)$, we also have $g(\bar{t}) > g(0)$. Now, from

$$f(x + \bar{t}(y - x)) = g(\bar{t}) \geqslant g(0) = f(x) \geqslant f(y),$$

we must have

$$f_x(x + \bar{t}(y - x))\Big[y - \Big(x + \bar{t}(y - x)\Big)\Big]$$
$$= (1 - \bar{t})f_x(x + \bar{t}(y - x))(y - x) \leqslant 0.$$

This implies

$$g'(\bar{t}) = f_x(x + \bar{t}(y - x))(y - x) \leqslant 0,$$

which is a contradiction to $g'(\bar{t}) > 0$, a contradiction. $\qquad\qquad\square$

We point out that although quasi-convex functions are very similar to convex functions, they are significantly different. Here is another interesting point. If $G_1 \subseteq G_2 \subseteq \mathbb{R}^n$ are two convex sets and $f : G_2 \to \mathbb{R}$ is convex. Then $f(\cdot)$ is also convex in the small set G_1. This property will not hold for quasi-convex functions (see one of exercise problem).

We now look at Problem (2.2), under quasi-convex conditions.

Theorem 4.6. *Let $G \subseteq \mathbb{R}^n$ be open and convex and $f, h^i : G \to \mathbb{R}$ be quasi-convex. Suppose $\bar{x} \in G$ such that*

$$A\bar{x} = b, \quad h(\bar{x}) \leqslant 0, \tag{4.8}$$

and for some $\bar{\lambda} \in \mathbb{R}^m$, $\bar{\mu} \in \mathbb{R}^\ell_+$, it holds that

$$f_x(\bar{x}) + \bar{\lambda}^\top A + \bar{\mu}^\top h_x(\bar{x}) = 0, \tag{4.9}$$

$$\bar{\mu}^\top h(\bar{x}) = 0. \tag{4.10}$$

Moreover, either

$$f_x(\bar{x}) \neq 0, \tag{4.11}$$

or $f(\cdot)$ is convex. Then $\bar{x}$ is an optimal solution of Problem (2.2).

Proof. Let

$$F = \{x \in \mathbb{R}^n \mid Ax = b, \ h(x) \leqslant 0\}.$$

Clearly, F is convex. By our assumption, for any $x \in F$,

$$f_x(\bar{x})(x - \bar{x}) = -\bar{\lambda}^\top A(x - \bar{x}) - \bar{\mu}^\top h_x(\bar{x})(x - \bar{x}) = -\sum_{i=1}^{\ell} \bar{\mu}_i h_x^i(\bar{x})(x - \bar{x}).$$

Now, if $h^i(\bar{x}) < 0$, we must have $\bar{\mu}_i = 0$, and correspondingly,

$$\bar{\mu}_i h_x^i(\bar{x})(x - \bar{x}) = 0.$$

On the other hand, if $h^i(\bar{x}) = 0$, by the convexity of F, $\bar{x} + t(x - \bar{x}) \in F$ for $t \in (0, 1)$. Thus,

$$h^i(\bar{x} + t(x - \bar{x})) \leqslant 0.$$

Consequently, (note $h^i(\bar{x}) = 0$)

$$0 \geqslant \frac{h^i(\bar{x} + t(x - \bar{x})) - h^i(\bar{x})}{t} \to h_x^i(\bar{x})(x - \bar{x}),$$

which leads to (since $\bar{\mu}_i \geqslant 0$)

$$f_x(\bar{x})(x - \bar{x}) = -\sum_{i=1}^{\ell} \bar{\mu}_i h_x^i(\bar{x})(x - \bar{x}) \geqslant 0. \tag{4.12}$$

Hence, if $f(\cdot)$ is convex, then for any $x \in F$,

$$f(x) - f(\bar{x}) \geqslant f_x(\bar{x})(x - \bar{x}) \geqslant 0,$$

showing that $\bar{x}$ is an optimal solution to Problem (2.2).

Next, if $f(\cdot)$ is quasi-convex and $f_x(\bar{x}) \neq 0$, then since G is open and $\bar{x} \in G$, there exists a $z \in G$ such that

$$f_x(\bar{x})(z - \bar{x}) > 0.$$

Now, pick any $x \in F$. By the convexity of G, for any $t \in (0, 1)$,

$$\xi(t) = (1 - t)x + tz \in G, \quad \eta(t) = (1 - t)\bar{x} + tz \in G.$$

Then

$$f_x(\bar{x})[\eta(t) - \bar{x}] = t f_x(\bar{x})(z - \bar{x}) > 0,$$

and by (4.12),

$$f_x(\bar{x})[\xi(t) - \eta(t)] = (1 - t)f_x(\bar{x})(x - \bar{x}) \geqslant 0.$$

Adding the above two together, we obtain

$$f_x(\bar{x})[\xi(t) - \bar{x}] > 0.$$

Then, by Theorem 4.5, we must have

$$f(\bar{x}) \leqslant f(\xi(t)), \qquad t \in (0,1),$$

since $f(\bar{x}) \geqslant f(\xi(t))$ implies $f_x(\bar{x})(\xi(t) - \bar{x}) \leqslant 0$. Finally, sending $t \to 0$, we have

$$f(\bar{x}) \leqslant f(x),$$

proving the minimality of $\bar{x}$. $\square$

Exercises

1. Let $f, g : \mathbb{R} \to \mathbb{R}$ be quasi-convex. Is $f \circ g$ quasi-convex? Why? Is $f(\cdot) + g(\cdot)$ quasi-convex? Why?

2. Let $f(x, y) = x^\alpha y^\beta$, for $x, y \geq 0$. Show that

(i) $f(\cdot, \cdot)$ is quasi-concave for any $\alpha, \beta \geqslant 0$;

(ii) $f(\cdot, \cdot)$ is neither convex nor concave if $\alpha + \beta > 1$.

3. Let $f(x, y) = -xy$. Show that $f(x, y)$ is quasi-convex on $\mathbb{R}_+^2$, but not on $\mathbb{R}^2$.

Chapter 5

Linear Programming

1   Standard and Canonical Forms

Let us recall the following LP problem which was introduced in Chapter 2:

$$\begin{cases} \min \quad c^\top x, \\ \text{subject to} \quad Ax = b, \quad x \geqslant 0, \end{cases} \tag{1.1}$$

where $A \in \mathbb{R}^{m \times n}$, $b \in \mathbb{R}^m_+$ and $c \in \mathbb{R}^n \setminus \{0\}$ are given. In order (1.1) to be non-trivial, we will keep the following assumption:

$$\operatorname{rank} A = m < n. \tag{1.2}$$

We call (1.1) a *standard form* LP problem. From (1.2) we see that $n \geqslant 2$. The case $n = 1$ is trivial.

Before going further, we claim that any linear programming problem can be transformed into the above standard form (1.1). To see this, we first note that, if we have a maximization problem, then replacing c by $-c$, we obtain an equivalent minimization problem. Second, the most general linear constraints have the following form:

$$\begin{cases} A_1 x = b_1, \qquad A_2 x \leqslant b_2, \qquad A_3 x \geqslant b_3, \\ x \equiv \begin{pmatrix} x_1 \\ x_2 \\ x_3 \end{pmatrix}, \qquad x_1 \geqslant 0, \quad x_2 \leqslant 0, \quad x_3 \text{ is unrestricted,} \end{cases}$$

for some $A_i \in \mathbb{R}^{m_i \times n}$, $b_i \in \mathbb{R}^{m_i}_+$, and $x_i \in \mathbb{R}^{n_i}$ $(1 \leqslant i \leqslant 3)$, with $n_1 + n_2 + n_3 = n$. Let $\widetilde{x}_2 = -x_2$ and introduce $x_4, x_5 \in \mathbb{R}^{n_3}_+$. Replacing x_2 by $-\widetilde{x}_2$ and x_3 by $x_4 - x_5$ in the above, we obtain that

$$\widetilde{A}_1 \widetilde{x} = b_1, \quad \widetilde{A}_2 \widetilde{x} \leqslant b_2, \quad \widetilde{A}_3 \widetilde{x} \geqslant b_3, \quad \widetilde{x} \equiv \begin{pmatrix} x_1 \\ \widetilde{x}_2 \\ x_4 \\ x_5 \end{pmatrix} \geqslant 0, \tag{1.3}$$

where

$$\widetilde{A}_i = A_i \begin{pmatrix} I & 0 & 0 & 0 \\ 0 & -I & 0 & 0 \\ 0 & 0 & I & -I \end{pmatrix}, \qquad i = 1, 2, 3.$$

Next, we introduce *slack variables* ξ_2, ξ_3, and replace (1.3) by

$$\begin{pmatrix} \widetilde{A}_1 & 0 & 0 \\ \widetilde{A}_2 & I & 0 \\ \widetilde{A}_3 & 0 & -I \end{pmatrix} \begin{pmatrix} \widetilde{x} \\ \xi_2 \\ \xi_3 \end{pmatrix} = \begin{pmatrix} b_1 \\ b_2 \\ b_3 \end{pmatrix}, \qquad \begin{pmatrix} \widetilde{x} \\ \xi_2 \\ \xi_3 \end{pmatrix} \geqslant 0,$$

which is a form of the constraint in (1.1).

From the above, we see that one can start with (1.1) under condition (1.2), without loss of generality. Also, we point out that for standard form LP problem (1.1), there is no loss of generality to assume that

$$b \geqslant 0, \tag{1.4}$$

since we can change the sign in each row of A if necessary.

Next, we consider the following LP problem:

$$\begin{cases} \min & c^\top x, \\ \text{subject to } Ax \geqslant b, \quad x \geqslant 0. \end{cases} \tag{1.5}$$

The above is called a *canonical form* LP problem. Sometimes it is more convenient to use such a form. The difference between (1.1) and (1.5) is that the equality constraint in (1.1) becomes an inequality constraint in (1.5). We have seen in the above that (1.5) can be put into the standard form (1.1) by some tranformations. On the other hand, $Ax = b$ is equivalent to the following:

$$\begin{pmatrix} A \\ -A \end{pmatrix} x \geqslant \begin{pmatrix} b \\ -b \end{pmatrix}.$$

However, unlike the standard form LP problem (1.1), for the canonical form (1.5), we could not assume (1.4) in general. Also, we cannot assume that $\operatorname{rank} A = m$.

For convenience, we introduce the following definition.

Definition 1.1. LP (1.1) is said to be *feasible* if the feasible set

$$F = \{ x \in \mathbb{R}^n \mid Ax = b, \quad x \geqslant 0 \}$$

is non-empty; LP problem (1.1) is said to be *infeasible* if $F = \varnothing$; LP problem (1.1) is said to be unbounded if $F \neq \varnothing$, and

$$\inf_{x \in F} c^\top x = -\infty.$$

We now look at some simple examples.

Example 1.2. Consider the following problem:

$$\begin{cases} \min & x_1 + x_2, \\ \text{subject to} & x_1 + x_2 \geqslant 1, \quad x_1, x_2 \geqslant 0. \end{cases}$$

Note that for any $x_1^*, x_2^* \geqslant 0$ satisfying $x_1^* + x_2^* = 1$, (x_1^*, x_2^*) is an optimal solution. Thus, optimal solution could be non-unique.

Example 1.3. Consider the following problem:

$$\begin{cases} \min & x_1 + 2x_2, \\ \text{subject to} & \begin{cases} x_1 + x_2 \geqslant 1, \\ x_1 + x_2 \leqslant 0, \\ x_1, x_2 \geqslant 0. \end{cases} \end{cases}$$

It is clear that the problem is infeasible. Therefore, there is no solution for the problem.

Example 1.4. Consider the following problem:

$$\begin{cases} \min & x_1 - 2x_2, \\ \text{subject to} & x_1 - x_2 \leqslant 0, \quad x_1, x_2 \geqslant 0. \end{cases}$$

It is easy to see that the problem is unbounded. Therefore, such a problem does not have a solution.

Let us now look at an LP problems in $\mathbb{R}^2$, which can be solved by some geometric arguments.

Example 1.5. Consider the following problem:

$$\begin{cases} \min & x_1 + x_2, \\ \text{subject to} & \begin{cases} x_1 + 2x_2 \geqslant 4, \\ -x_1 + x_2 \leqslant 1, \\ 2x_1 + x_2 \leqslant 4, \\ x_1, x_2 \geqslant 0. \end{cases} \end{cases}$$

For this problem, the feasible set F is a region bounded by the straight lines: $x_1 + 2x_2 = 4$, $-x_1 + x_2 = 1$, and $2x_1 + x_2 = 4$ in the first quadrant. This is a triangle with the vertices:

$$\left(\frac{2}{3}, \frac{5}{3}\right), \quad \left(\frac{4}{3}, \frac{4}{3}\right), \quad (1, 2).$$

Now, for any $\alpha \in \mathbb{R}$, $x_1 + x_2 = \alpha$ is a straight line with normal direction $(1, 1)$. Moving in this direction, the objective function is increasing. To minimize $x_1 + x_2$, one needs to move this line in the direction $-(1, 1)$, keeping the line intersecting F. The optimal solution is given by the "last point(s)" in F that the line intersects F. Thus, the optimal solution is given by $(\frac{2}{3}, \frac{5}{3})$, with the minimum objective function value $\frac{7}{3}$. One may also directly evaluate the objective function at these three vertices, and choose the one with the smallest objective function value (we will see that this must be the optimal solution). Likewise, if we want to maximize $x_1 + x_2$, then we move the line in the direction $(1, 1)$ and the maximum is the "last point(s)" in F that the line intersects F. Hence, the maximum is attained at $(1, 2)$, with the maximum objective function value 3.

The above example illustrates a method that works for LP problems in $\mathbb{R}^2$, when the constraints are not very complicated.

Exercises

1. Transform the following into standard form:
$$\begin{cases} \min & 2x_1 - x_2 + 3x_3, \\ \text{subject to} & \begin{cases} x_1 - 2x_2 = 3, \\ -2x_2 + 3x_3 \leqslant 0, \\ 3x_2 + x_3 \leqslant -2, \\ x_1, x_2 \geqslant 0, \quad x_3 \leqslant 0. \end{cases} \end{cases}$$

2. Transform the following into canonical form:
$$\begin{cases} \max & x_1 + 2_2 - 3x_3, \\ \text{subject to} & \begin{cases} 2x_1 + 2x_2 = 3, \\ x_1 - 2x_2 + x_3 \leqslant 0, \\ 3x_2 - x_3 \geqslant -2, \\ x_1 \geqslant 0, \quad x_2, x_3 \leqslant 0. \end{cases} \end{cases}$$

3. Consider standard form LP problem (1.1). Let F be the corresponding feasible set. Show that F is convex and closed. Do the same thing for the canonical form LP problem (1.5).

4. Mimicking Example 1.5 to solve the following LP problem:
$$\begin{cases} \min & x_1 + 2x_2, \\ \text{subject to} & 0 \leqslant x_1 \leqslant 4, \quad 0 \leqslant x_2 \leqslant x_1 + 3, \quad 3x_1 + x_2 \leqslant 9. \end{cases}$$

2 Geometric Considerations and the Fundamental Theorem of Linear Programming

In this section, we study LP from geometric point of view. In the rest of this section, we denote the *feasible set F* of LP (1.1) by the following:

$$F = \{x \in \mathbb{R}^n \mid Ax = b, \ x \geq 0\}. \tag{2.1}$$

Our first result is the following.

Proposition 2.1. *Suppose LP (1.1) admits an optimal solution $\bar{x} \in F$. Then it is necessary that $\bar{x} \in \partial F$.*

Proof. Suppose the objective function $f(x) \equiv c^\top x$ attains a local minimum at $\bar{x} \in F^\circ$. Then by Fermat's theorem,

$$f_x(\bar{x}) = c^\top = 0,$$

which is a contradiction since $c \neq 0$. Hence, our conclusion follows. □

Note that LP problem (1.1) can be regarded as an optimization problem with equality and inequality constraints. Therefore, we can try to use the KKT method to solve it. Here is such a result.

Proposition 2.2. *For LP problem (1.1), an $x^* \in F$ is an optimal solution if there exist $\lambda \in \mathbb{R}^m$ and $\mu \in \mathbb{R}^n$ such that*

$$\begin{cases} c^\top + \lambda^\top A - \mu^\top = 0, & \mu \geq 0, \\ Ax^* = b, & \mu_i x_i^* = 0, \quad 1 \leq i \leq n. \end{cases} \tag{2.2}$$

Proof. Suppose we have (2.2). Then for any $x \in F$,

$$c^\top x - c^\top x^* = (\mu^\top - \lambda^\top A)x - (\mu^\top - \lambda^\top A)x^*$$
$$= \mu^\top x - \lambda^\top Ax - \mu^\top x^* + \lambda^\top Ax^* = \mu^\top x \geq 0.$$

Hence, x^* is optimal. □

Note that in the case that the constraints are regular at an optimal solution x^*, (2.2) is just the first order KKT conditions.

Let us present an example.

Example 2.3. Consider the following LP problem:

$$\begin{cases} \min & x_1 - x_2, \\ \text{subject to} & x_1 + 2x_2 = 1, \quad x_1, x_2 \geq 0. \end{cases}$$

In the current case, $m = 1$, $n = 2$, and

$$A = (1, 2), \quad b = 1, \quad c^\top = (1, -1).$$

Then we need to solve the following:

$$\begin{cases} x_1 + 2x_2 = 1, \\ \mu_1 x_1 = \mu_2 x_2 = 0, \\ 1 + \lambda - \mu_1 = 0, \\ -1 + 2\lambda - \mu_2 = 0, \\ x_1, x_2, \mu_1, \mu_2 \geqslant 0. \end{cases}$$

Thus,

$$x_1 = 1 - 2x_2, \quad \mu_1 = \lambda + 1 = \frac{\mu_2 + 1}{2} + 1 = \frac{\mu_2 + 3}{2} > 0.$$

Then

$$x_1 = 0, \quad x_2 = \frac{1}{2}, \quad \mu_2 = 0.$$

Hence,

$$\mu_1 = \frac{3}{2}, \quad \lambda = \frac{1}{2}.$$

By Proposition 2.2, $(x_1, x_2) = (0, \frac{1}{2})$ is optimal.

We note that using Proposition 2.2 to solve LP problems, in general, there are $2n + m$ unknowns $x_1, \cdots, x_n$, $\mu_1, \cdots, \mu_n$, $\lambda_1, \cdots, \lambda_m$, and there is a system of $2n + m$ equations plus the requirement that $x, \mu \geqslant 0$. In principle, we can solve LP in this way. However, when the number of decision variables is big, the above method does not lead to an efficient algorithm. Therefore, we would like to look at the LP problems more directly.

Definition 2.4. A set $P \subseteq \mathbb{R}^n$ is called a *polyhedron* if it is the non-empty intersection of a finite set of closed half spaces. When this set is bounded, we further call it a *polytope*.

Any polyhedron P admits the following representation:

$$P = \left\{ x \in \mathbb{R}^n \mid a_i^\top x \leqslant \beta_i, \ 1 \leqslant i \leqslant k \right\},$$

for some $a_i \in \mathbb{R}^n \setminus \{0\}$ and $\beta_i \in \mathbb{R}$ $(i = 1, 2, \cdots, k)$. Clearly, the feasible set F of an LP is a polyhedron. It should be pointed out that a polyhedron in $\mathbb{R}^n$ does not necessarily to have interior points (in $\mathbb{R}^n$).

Definition 2.5. Let $G \subseteq \mathbb{R}^n$ be convex. A point $x \in \partial G$ is called an *extreme point* of G if it is not a convex combination of any two distinct points in G, i.e.,

$$x = \lambda y + (1 - \lambda)z, \quad y, z \in G, \ 0 < \lambda < 1,$$

implies $x = y = z$.

We will denote the set of all extreme points of G by $\mathcal{E}(G)$. For a polyhedron P, any point $x \in \mathcal{E}(G)$ is called a *vertex* of P.

Now, we return to our standard form of LP (1.1). The following result characterizes the extreme points of the feasible set F, defined by (2.1), of LP (1.1).

Theorem 2.6. (i) *If* $0 \in F$, *then* $0 \in \mathcal{E}(F)$.

(ii) *A non-zero point* $x^* \in \mathcal{E}(F)$ *if and only if the columns of A corresponding to the positive components of x^* are linearly independent. Consequently, any* $x^* \in \mathcal{E}(F)$ *has at most m positive components.*

(iii) *If the feasible set F is non-empty, then $\mathcal{E}(F)$ is non-empty.*

Proof. (i) If $0 \in F$ and it is not an extreme point, then there are $x, y \in F$ such that $x \neq y$ and for some $\lambda \in (0, 1)$,

$$\lambda x + (1 - \lambda)y = 0.$$

Looking at the components of x and y, which are supposed to be nonnegative, we see that it is necessarily $x = y = 0$, a contradiction.

(ii) Without loss of generality (by relabelling the components of the decision variable and correspondingly exchanging the columns of A), we may assume that x^* has the following form:

$$x^* = \begin{pmatrix} \bar{x} \\ 0 \end{pmatrix}, \qquad \bar{x} \equiv (x_1, \cdots, x_k)^\top > 0.$$

Accordingly, we let

$$A = (\bar{A}, \tilde{A}), \qquad \bar{A} \in \mathbb{R}^{m \times k}, \ \tilde{A} \in \mathbb{R}^{m \times (n-k)}.$$

Then, by the equality constraint, we have

$$\bar{A}\bar{x} = Ax^* = b.$$

Now, if the columns of $\bar{A}$ are linearly dependent, then there exists a $\bar{y} \in \mathbb{R}^k \setminus \{0\}$ such that

$$\bar{A}\bar{y} = 0.$$

Since $\bar{x} > 0$, we can find a $\delta > 0$ such that

$$\bar{x} + \delta\bar{y} \geqslant 0.$$

Hence, we see that

$$z^{\pm} \triangleq x^* \pm \begin{pmatrix} \bar{y} \\ 0 \end{pmatrix} \in F,$$

and

$$x^* = \frac{1}{2}z^+ + \frac{1}{2}z^-.$$

Since $z^+ \neq z^-$, we see that $x^* \notin \mathcal{E}(F)$.

Conversely, if $x^* \in F \setminus \mathcal{E}(F)$, then we can find $y, z \in F$, $y \neq z$, and $0 < \lambda < 1$ such that

$$x^* = \lambda y + (1 - \lambda)z.$$

Since the last $n - k$ components are zero and $y, z \geqslant 0$, we must have

$$y = \begin{pmatrix} \bar{y} \\ 0 \end{pmatrix}, \qquad z = \begin{pmatrix} \bar{z} \\ 0 \end{pmatrix},$$

and $\bar{y} \neq \bar{z}$. Hence,

$$\bar{A}(\bar{y} - \bar{z}) = A(y - z) = b - b = 0.$$

This means that the columns of $\bar{A}$ are linearly dependent.

Finally, since A has rank m, we must have $k \leqslant m$.

(iii) Let $x \in F$. Since $x \geqslant 0$, without loss of generality, we assume that

$$x = \begin{pmatrix} \bar{x} \\ 0 \end{pmatrix}, \qquad \bar{x} = (x_1, x_2, \cdots, x_k)^{\top} > 0.$$

Let $A = (a_1, a_2, \cdots, a_n)$ with $a_i \in \mathbb{R}^m$, $1 \leqslant i \leqslant n$. If the columns $a_1, \cdots, a_k$ are linearly independent, then by (ii) of this proposition, $x \in \mathcal{E}(F)$, and we are done. Suppose otherwise, then there is a $\bar{y} = (y_1, y_2, \cdots, y_k) \neq 0$ such that

$$0 = \sum_{i=1}^{k} y_i a_i = (a_1, a_2, \cdots, a_k)\bar{y} \equiv Ay,$$

where $y = \begin{pmatrix} \bar{y} \\ 0 \end{pmatrix}$. Next, let

$$\min\left\{ \frac{x_i}{|y_i|} \,\Big|\, 1 \leqslant i \leqslant k, \; y_i \neq 0 \right\} = \frac{x_{i_0}}{|y_{i_0}|} > 0$$

for some $i_0 \in \{1, 2, \cdots, k\}$, and let

$$\widehat{x} = x - \frac{x_{i_0}}{y_{i_0}} y.$$

Then we have

$$A\widehat{x} = Ax - \frac{x_{i_0}}{y_{i_0}} Ay = Ax = b,$$

and by the definition of i_0, one has

$$\widehat{x} \geqslant 0, \quad \widehat{x}_{i_0} = \widehat{x}_{k+1} = \cdots \widehat{x}_n = 0.$$

Thus, we find an $\widehat{x} \in F$ such that the number of positive components is smaller than k. Continuing this procedure, we will end up with an $x^* \in F$ such that the column vectors of A corresponding to the positive components of x^* are linearly independent. Hence, $x^* \in \mathcal{E}(F)$, proving the non-emptiness of $\mathcal{E}(F)$. $\qquad\qquad\square$

Let us present an example to illustrate the above result.

Example 2.7. Let

$$A = \begin{pmatrix} 1 & 2 & 3 \\ 0 & -2 & 1 \end{pmatrix}, \quad b = \begin{pmatrix} 1 \\ 0 \end{pmatrix}.$$

We now find the feasible set F. To this end, we solve the following system of equations:

$$\begin{cases} x_1 + 2x_2 + 3x_3 = 1, \\ \quad\;\; -2x_2 + x_3 = 0. \end{cases}$$

The general solution is given by

$$x = \begin{pmatrix} 1 \\ 0 \\ 0 \end{pmatrix} + \begin{pmatrix} -8 \\ 1 \\ 2 \end{pmatrix} x_2.$$

Thus,

$$F = \left\{ \begin{pmatrix} 1 \\ 0 \\ 0 \end{pmatrix} + \begin{pmatrix} -8 \\ 1 \\ 2 \end{pmatrix} x_2 \;\Big|\; 0 \leqslant x_2 \leqslant \frac{1}{8} \right\}.$$

Let us look at the following two points (corresponding to $x_2 = 0, \frac{1}{8}$):

$$x^* = \begin{pmatrix} 1 \\ 0 \\ 0 \end{pmatrix}, \qquad x^{**} = \begin{pmatrix} 0 \\ 1 \\ 2 \end{pmatrix}.$$

For x^*, the column of A corresponding to the positive component of x^* is just the first column of A, which is linearly independent (since it is non-zero). For x^{**}, the columns of A corresponding to the positive components of x^{**} are the last two columns of A, which are linearly independent. On the other hand, with $0 < x_2 < \frac{1}{8}$, the corresponding solution has all three components strictly positive. The corresponding columns of A are the all three columns which are linearly dependent. Hence, such kind of points are not extreme points of F. Consequently,

$$\mathcal{E}(F) = \{x^*, x^{**}\}.$$

The following is a weak version of *Fundamental Theorem of Linear Programming*.

Theorem 2.8. *For the standard LP problem* (1.1), *let its feasible set*

$$F = \{x \in \mathbb{R}^n \mid Ax = b, x \geqslant 0\}$$

be non-empty and $\mathcal{E}(F)$ be the set of extreme points of F. If the LP problem has an optimal solution, then there exists an optimal solution $x^* \in \mathcal{E}(F)$.

Proof. Let $x^* \in F$ be an optimal solution to (1.1). If $x^* \in \mathcal{E}(F)$, we are done. Suppose $x^* \in F \setminus \mathcal{E}(F)$. Without loss of generality, we assume that

$$x^* = \begin{pmatrix} \bar{x} \\ 0 \end{pmatrix}, \quad \bar{x} = (\bar{x}_1, \cdots, \bar{x}_k)^\top \in \mathbb{R}^k, \quad \bar{x} > 0.$$

Correspondingly, we decompose

$$A = (\bar{A}, \tilde{A}), \qquad \bar{A} \in \mathbb{R}^{m \times k}, \quad \tilde{A} \in \mathbb{R}^{m \times (n-k)}.$$

By Theorem 2.6, we know that the columns of $\bar{A}$ are linearly dependent. Therefore, there exists a $\bar{y} = (\bar{y}_1, \cdots, \bar{y}_k)^\top \in \mathbb{R}^k \setminus \{0\}$ such that

$$\bar{A}\bar{y} = 0.$$

There will be an i_0 with $1 \leqslant i_0 \leqslant k$ such that

$$\frac{\bar{x}_{i_0}}{|\bar{y}_{i_0}|} = \min\left\{ \frac{\bar{x}_i}{|\bar{y}_i|} \mid 1 \leqslant i \leqslant k, \ y_i \neq 0 \right\} \equiv \bar{\delta} > 0.$$

Then for any $\delta \in [-\bar{\delta}, \bar{\delta}]$,

$$x(\delta) = x^* + \delta y \in F,$$

where $y = \begin{pmatrix} \bar{y} \\ 0 \end{pmatrix}$. Moreover,

$$\hat{x} \equiv x\left(\frac{\bar{x}_{i_0}}{\bar{y}_{i_0}}\right) = x^* - \frac{\bar{x}_{i_0}}{\bar{y}_{i_0}} y$$

has at most $k - 1$ positive components. Furthermore, by the optimality of x^*, one has

$$c^\top x^* \leqslant c^\top x(\delta) = c^\top x^* + \delta c^\top y, \qquad \forall \delta \in [-\bar{\delta}, \bar{\delta}].$$

Hence, it is necessary that

$$c^\top y = 0.$$

Consequently, we obtain

$$c^\top \hat{x} = c^\top x^* = \min_{x \in F} c^\top x.$$

Therefore, we find another minimum $\hat{x} \in F$, which has less positive components than x^*. Continue this procedure, we will end up with our conclusion. $\qquad\blacksquare$

In principle, the above result tells us that to solve our LP problem, it suffices to find all extreme points $\mathcal{E}(F)$ of the feasible set F, and then evaluate the objective function $c^\top x$ at these extreme points to obtain the optimal solution (either maximum or minimum). When the number of constraints is not large, and the feasible set F is compact, this is actually working. We now present an example.

Example 2.9. We consider the following problem:

$$\begin{cases} \min & x_1 + 2x_2 + 3x_3 + 4x_4, \\ \text{subject to} & \begin{cases} 2x_1 - 3x_2 + 5x_4 = 0, \\ x_1 + 3x_2 - 2x_4 = 9, \\ -x_2 + x_3 + 3x_4 = 2, \\ x_1, x_2, x_3, x_4 \geqslant 0. \end{cases} \end{cases}$$

In the current case, we have

$$A = \begin{pmatrix} 2 & -3 & 0 & 5 \\ 1 & 3 & 0 & -2 \\ 0 & -1 & 1 & 3 \end{pmatrix}, \quad b = \begin{pmatrix} 0 \\ 9 \\ 2 \end{pmatrix}, \quad c = \begin{pmatrix} 1 \\ 2 \\ 3 \\ 4 \end{pmatrix}.$$

We solve linear system

$$\begin{cases} 2x_1 - 3x_2 + 5x_4 = 0, \\ x_1 + 3x_2 - 2x_4 = 9, \\ -x_2 + x_3 + 3x_4 = 2. \end{cases}$$

The solution is given by the following:

$$\begin{cases} x_1 = 3 - x_4, \\ x_2 = 2 + x_4, \\ x_3 = 4 - 2x_4. \end{cases}$$

Thus, the solution is feasible if and only if in the above,

$$0 \leqslant x_4 \leqslant 2.$$

Hence, the feasible set F is given by

$$F = \left\{ \begin{pmatrix} 3 - x_4 \\ 2 + x_4 \\ 4 - 2x_4 \\ x_4 \end{pmatrix} \;\middle|\; x_4 \in [0, 2] \right\},$$

which is a compact set.

For $x_4 = 0$, we get a solution

$$x^* = \begin{pmatrix} 3 \\ 2 \\ 4 \\ 0 \end{pmatrix}.$$

The columns of A corresponding the positive components of x^* are the first three columns which are linearly independent. Thus, x^* is an extreme point of F.

For $x_4 = 2$, we get a solution

$$x^{**} = \begin{pmatrix} 1 \\ 4 \\ 0 \\ 2 \end{pmatrix}.$$

The columns of A corresponding the positive components of x^{**} are the first, the second and the fourth columns. These three columns are linearly independent. Thus, x^{**} is an extreme point of F also.

For other values of x_4, all the components of the solution are positive and the corresponding columns of A are linearly dependent. Hence. they are not extreme points of F. Thus,

$$\mathcal{E}(F) = \{x^*, x^{**}\}.$$

Now,

$$c^\top x^* = 3 + 4 + 12 = 19, \quad c^\top x^{**} = 1 + 8 + 8 = 17.$$

Hence, the minimum is attained at x^{**}.

We should be a little careful that if F is not compact, the above method might be misleading. Here is an example.

Example 2.10. Consider LP problem:

$$\begin{cases} \min & x_1 - x_2, \\ \text{subject to} & \begin{cases} x_1 + x_2 - x_3 = 1, \\ x_1, x_2, x_3 \geqslant 0. \end{cases} \end{cases}$$

In this case, $(x_1, x_2, x_3) \in F$ if and only if

$$x_3 = x_1 + x_2 - 1, \quad x_1, x_2, x_3 \geqslant 0.$$

Thus,

$$F = \left\{ \begin{pmatrix} x_1 \\ x_2 \\ x_1 + x_2 - 1 \end{pmatrix} \;\middle|\; x_1, x_2 \geqslant 0, \; x_1 + x_2 \geqslant 1 \right\},$$

which is not compact. But, we may still find extreme points as follows. In the current case, $m = 1$, $n = 3$, and

$$A = (1, 2, -1), \quad b = 1, \quad c^\top = (1, -1, 0).$$

Thus, by Theorem 2.6, we see that

$$x^* = \begin{pmatrix} 1 \\ 0 \\ 0 \end{pmatrix} \equiv e_1, \quad x^{**} = \begin{pmatrix} 0 \\ 1 \\ 0 \end{pmatrix} \equiv e_2$$

are the only extreme points of F. We have

$$c^\top x^* = 1, \quad c^\top x^{**} = -1.$$

But

$$\inf_{x \in F} c^\top x = \inf_{x_1, x_2 \geqslant 0} (x_1 - x_2) = -\infty.$$

Hence, x^{**} is not a solution of LP problem.

We now modify the above example, which will lead to an interesting question.

Example 2.11. Consider

$$
\begin{cases}
\min & x_1 + 2x_2, \\
\text{subject to} & \begin{cases} x_1 + x_2 - x_3 = 1, \\ x_1, x_2, x_3 \geqslant 0. \end{cases}
\end{cases}
$$

The same as Example 2.10, we still have unbounded feasible set F and

$$\mathcal{E}(F) = \{e_1, e_2\}.$$

But,

$$\inf_{x \in F} c^\top x = \inf_{x \in F} (x_1 + 2x_2) \geqslant 0.$$

Can we say that for this example, one has an optimal solution?

In general, we can ask the following question: Suppose F is non-empty and unbounded. Suppose that

$$\inf_{x \in F} c^\top x > -\infty.$$

Is there an optimal solution to the corresponding LP problem? Note that for nonlinear functions, one cannot expect that, for example, $f(x) = e^{-x^2}$.

To answer this question, we introduce the following notion for our LP problem.

Definition 2.12. A vector $d \in \mathbb{R}^n$ is called an *extreme direction* of a feasible set F if for some $x \in F$,

$$x + \lambda d \in F, \qquad \forall \lambda \geqslant 0.$$

Note that zero vector is a (trivial) extreme direction. By accepting 0 as an extreme direction, it will make some statements below simpler. It is easy to see that any bounded feasible set does not have non-zero extreme directions. Now, for the feasible set F of LP (1.1), we have the following simple proposition.

Proposition 2.13. (i) *A vector $d \in \mathbb{R}^n$ is an extreme direction of F if and only if*

$$Ad = 0, \qquad d \geqslant 0. \tag{2.3}$$

(ii) *Feasible set F is unbounded if and only if it admits a non-zero extreme direction.*

Proof. (i) By Definition 2.12, $d \in \mathbb{R}^n$ is an extreme direction of F if and only if for some $x \in F$,

$$A(x + \lambda d) = b + \lambda Ad = b, \qquad x + \lambda d \geqslant 0, \qquad \forall \lambda \geqslant 0.$$

This is clearly equivalent to (2.3).

(ii) If F admits a non-zero extreme direction, then F clearly is unbounded. Conversely, if F is unbounded, then there exists a sequence $x^k \in F$ such that $\|x^k\| \to \infty$ as $k \to \infty$. Let

$$d^k = \frac{x^k}{\|x^k\|}, \qquad k \geqslant 1.$$

Then d^k lies on the unit sphere. Hence, we can find a convergent subsequence. Without loss of generality, we assume that

$$\lim_{k \to \infty} \|d^k - d\| = 0.$$

Clearly, $\|d\| = 1$ and $d \geqslant 0$, also,

$$Ad = \lim_{k \to \infty} A \frac{x^k}{\|x^k\|} = \lim_{k \to \infty} \frac{b}{\|x^k\|} = 0.$$

Hence, by (i), d is a non-zero extreme direction of F. $\qquad\square$

The above result tells us that the extreme directions of a feasible set F (determined by some A and b) is completely determined by A only.

For the above Example 2.7 by solving

$$Ad = 0,$$

we get that

$$d = \begin{pmatrix} -8 \\ 1 \\ 2 \end{pmatrix} x_2, \qquad x_2 \geqslant 0.$$

Hence, for this example, F does not have non-trivial extreme directions. This is actually a case that the feasible set F is bounded. We now look at another example.

Example 2.14. Let

$$A = \begin{pmatrix} 1 & 2 & -5 \\ 2 & -1 & 0 \end{pmatrix}, \qquad b = \begin{pmatrix} 0 \\ 1 \end{pmatrix}.$$

Solving the homogeneous equation:

$$\begin{cases} x_1 + 2x_2 - 5x_3 = 0, \\ 2x_1 - x_2 = 0. \end{cases}$$

Thus, we have general solution

$$x = \begin{pmatrix} 1 \\ 2 \\ 1 \end{pmatrix} x_1.$$

For this example, we see that $d = (1, 2, 1)^\top$ is an extreme direction of F.

The following gives a representation of the feasible set F in terms of extreme point set $\mathcal{E}(F)$ of F and extreme directions of F.

Theorem 2.15. *For LP* (1.1), *let* $\mathcal{E}(F) = \{v^i \in \mathbb{R}^n \mid i \in \Lambda\}$. *Then any* $x \in F$ *admits a representation*

$$x = \sum_{i \in \Lambda} \lambda_i v^i + d, \tag{2.4}$$

where

$$\sum_{i \in \Lambda} \lambda_i = 1, \qquad \lambda_i \geqslant 0, \quad i \in \Lambda, \tag{2.5}$$

and d *is some extreme direction of* F.

Proof. Let $x \in F$. We use induction on the number k of the positive components of x. If $k = 0$, then $x = 0$, which is the trivial extreme point of F. Hence, (2.4) holds. Now, suppose (2.5) holds for all $x \in F$ that have no more than k positive components. Let $x \in F$ have exactly $(k + 1)$ positive components. Without loss of generality, we let

$$x = \begin{pmatrix} \bar{x} \\ 0 \end{pmatrix}, \qquad \bar{x} = (x_1, \cdots, x_{k+1})^\top > 0.$$

Correspondingly, we let

$$A = (\bar{A}, \widetilde{A}), \qquad \bar{A} \in \mathbb{R}^{m \times (k+1)}, \quad \widetilde{A} \in \mathbb{R}^{m \times (n-k-1)}.$$

If x is an extreme point of F, we are done. Now, suppose x is not an extreme point. Then by Theorem 2.6, we know that the columns of $\bar{A}$ are linearly dependent. Hence, we can find a $\bar{y} \equiv (y_1, \cdots, y_{k+1})^\top \in \mathbb{R}^{k+1} \setminus \{0\}$, such that

$$\bar{A}\bar{y} = 0.$$

Define

$$y = \begin{pmatrix} \bar{y} \\ 0 \end{pmatrix} \in \mathbb{R}^n.$$

Then $y \neq 0$ and

$$Ay = \bar{A}\bar{y} = 0.$$

We consider three cases.

(i) $\bar{y} \geq 0$. Without loss of generality, we assume that $y_1, y_2, \cdots, y_\ell > 0$, and $y_{\ell+1} = \cdots = y_n = 0$, with $\ell \leq k+1$. Also, we may assume that

$$\frac{x_1}{y_1} = \min\left\{\frac{x_i}{y_i} \;\middle|\; 1 \leq i \leq \ell\right\} \equiv \theta > 0.$$

Define

$$z \equiv x - \theta y.$$

Then

$$\begin{cases} z_i = x_i - \theta y_i = \left(\dfrac{x_i}{y_i} - \theta\right) y_i \geq 0, & 1 \leq i \leq \ell, \\ z_j = x_j \geq 0, & \ell+1 \leq j \leq n. \end{cases}$$

Therefore,

$$Az = b, \quad z \geq 0, \quad z_1 = x_1 - \theta y_1 = 0, \quad z_{k+2} = \cdots = z_n = 0.$$

We see that z has at most k positive components. Hence, by induction hypothesis, we have

$$z = \sum_{i \in \Lambda} \lambda_i v^i + d, \tag{2.6}$$

with λ_i satisfying (2.5) and d being an extreme direction of F. Consequently,

$$x = z + \theta y = \sum_{i \in \Lambda} \lambda_i v^i + (d + \theta y).$$

By Proposition 2.13 (i), $d + \theta y \geq 0$ is an extreme direction of F. Thus, (2.4) holds for x in this case.

(ii) $\bar{y} \leq 0$. Without loss of generality, we assume that $y_1, y_2, \cdots, y_\ell < 0$, and $y_{\ell+1} = \cdots = y_n = 0$, with $\ell \leq k+1$. Also, we may assume that

$$\frac{x_1}{-y_1} = \min\left\{\frac{x_i}{-y_i} \;\middle|\; 1 \leq i \leq \ell\right\} \equiv \theta > 0.$$

Define

$$z \equiv x + \theta y.$$

Then

$$\begin{cases} z_i = x_i + \theta y_i = \left(\dfrac{x_i}{-y_i} - \theta \right)(-y_i) \geqslant 0, & 1 \leqslant i \leqslant \ell, \\ z_j = x_i \geqslant 0, & \ell + 1 \leqslant j \leqslant n. \end{cases}$$

Therefore,

$$Az = b, \quad z \geqslant 0, \quad z_1 = x_1 + \theta y_1 = 0, \quad z_{k+2} = \cdots = z_n = 0.$$

From the above, we see that z has at most k positive components. Hence, by induction hypothesis, we have

$$z = \sum_{i \in \Lambda} \lambda_i v^i + d, \tag{2.7}$$

with λ_i satisfying (2.5) and d being an extreme direction of F. Consequently,

$$x = z - \theta y = \sum_{i \in \Lambda} \lambda_i v^i + (d - \theta y).$$

By Proposition 2.13 (i) again, $d - \theta y \geqslant 0$ is an extreme direction of F. Thus, (2.4) holds for x in this case.

(iii) $\bar{y}$ has positive and negative components. Without loss of generality, we assume that

$$\bar{y} = (y_1, \cdots, y_\ell, y_{\ell+1}, \cdots, y_{k+1})^\top,$$

with

$$y_1, y_2, \cdots, y_\ell > 0, \quad y_{\ell+1}, y_{\ell+2}, \cdots, y_{\ell+\bar{\ell}} < 0, \quad \ell + \bar{\ell} \leqslant k + 1.$$

Then we can assume that

$$\frac{x_1}{y_1} = \min \left\{ \frac{x_i}{y_i} \;\middle|\; 1 \leqslant i \leqslant \ell \right\} \equiv \theta > 0.$$

Define

$$z = x - \theta y.$$

Then similar to case (i), we have

$$Az = b, \quad z \geqslant 0, \quad z_1 = x_1 - \frac{x_1}{y_1} y_1 = 0, \quad z_{k+2} = \cdots = z_n = 0.$$

Hence, by induction hypothesis, we obtain (the same as (2.7)).

$$x - \theta y = z = \sum_{i \in \Lambda} \mu_i v^i + d,$$

for some extreme direction d of F and with

$$\mu_i \geqslant 0, \qquad \sum_{i=1}^{\ell} \mu_i = 1.$$

Next, we can also assume that

$$\frac{x_{\ell+1}}{-y_{\ell+1}} = \min\left\{ \frac{x_i}{-y_i} \,\Big|\, \ell + 1 \leqslant i \leqslant \ell + \bar{\ell} \right\} \equiv \bar{\theta} > 0.$$

Define

$$\bar{z} = x + \bar{\theta} y.$$

Then similar to case (ii), we have

$$A\bar{z} = b, \quad \bar{z} \geqslant 0, \quad \bar{z}_{\ell+1} = x_{\ell+1} + \frac{x_{\ell+1}}{-y_{\ell+1}} y_{\ell+1} = 0, \quad \bar{z}_{k+2} = \cdots = \bar{z}_n = 0.$$

Hence, by induction hypothesis, we obtain

$$x + \bar{\theta} y = \bar{z} = \sum_{i \in \Lambda} \bar{\mu}_i v^i + \bar{d},$$

for some extreme direction $\bar{d}$ of F and with

$$\bar{\mu}_i \geqslant 0, \qquad \sum_{i \in \Lambda} \bar{\mu}_i = 1.$$

Then

$$(\theta + \bar{\theta})x = \theta(\bar{z} - \bar{\theta} y) + \bar{\theta}(z + \theta y)$$
$$= \theta \bar{z} + \bar{\theta} z = \sum_{i \in \Lambda} (\theta \bar{\mu}_i + \bar{\theta} \mu_i) v^i + \theta \bar{d} + \bar{\theta} d.$$

Consequently,

$$x = \sum_{i \in \Lambda} \frac{\theta \bar{\mu}_i + \bar{\theta}_i \mu_i}{\theta + \bar{\theta}} v^i + \frac{\theta \bar{d} + \bar{\theta} d}{\theta + \bar{\theta}} \equiv \sum_{i \in \Lambda} \widehat{\lambda}_i v^i + \widehat{d}.$$

By Proposition 2.13 (i), we see that $\widehat{d} \equiv \frac{\theta \bar{d} + \bar{\theta} d}{\theta + \bar{\theta}} \geqslant 0$ is an extreme direction of F, and

$$\widehat{\lambda}_i \equiv \frac{\theta \bar{\mu}_i + \bar{\theta} \mu_i}{\theta + \bar{\theta}} \geqslant 0, \quad \sum_{i \in \Lambda} \widehat{\lambda}_i = 1.$$

This completes our induction. $\qquad\qquad\qquad\qquad\qquad\qquad\qquad\square$

Now we state and prove the following strong version of the *Fundamental Theorem of Linear Programming*, comparing with the weak version, Theorem 2.8.

Theorem 2.16. *Let LP* (1.1) *be given such that the feasible set $F \neq \varnothing$. Then either*

$$\inf_{x \in F} c^{\top}x = -\infty, \tag{2.8}$$

or there exists a $v \in \mathcal{E}(F)$ such that

$$\min_{x \in F} c^{\top}x = c^{\top}v. \tag{2.9}$$

Proof. Let $\mathcal{E}(F) = \{v^i \mid i \in \Lambda\}$. We look at two mutually exclusive cases:

(i) There exists an extreme direction $d \in \mathbb{R}^n$ of F such that

$$c^{\top}d < 0.$$

In this case, since $x + \lambda d \in F$ for some $x \in F$ and $\lambda \geqslant 0$, we see that (2.8) holds.

(ii) If the above (i) fails, then, by Theorem 2.15, for any $x \in F$, we have

$$x = \sum_{i \in \Lambda} \lambda_i v^i + d, \qquad \sum_{i \in \Lambda} \lambda_i = 1, \ \lambda_i \geqslant 0, \quad i \in \Lambda,$$

for some extreme direction $d \in \mathbb{R}^n$ with $c^{\top}d \geqslant 0$. Since $\mathcal{E}(F)$ is a finite set, we can find a $v \in \mathcal{E}(F)$ such that

$$c^{\top}v = \min_{i \in \Lambda} c^{\top}v^i.$$

Then

$$c^{\top}x \geqslant \sum_{i \in \Lambda} \lambda_i c^{\top}v^i \geqslant c^{\top}v \sum_{i \in \Lambda} \lambda_i = c^{\top}v.$$

Hence, (2.9) follows. ∎

Now, we can finish Example 2.11. Since the objective stays non-negative, by Theorem 2.16, the optimal solution must exists and must be at one of the extreme point. Hence, by evaluating

$$c^{\top}e_1 = 1, \qquad c^{\top}e_2 = 2,$$

we see that the minimum is attained at $e_1 = (1, 0, 0)^{\top}$.

Let us present another example.

Example 2.17. Consider the following linear programming problem:

$$\begin{cases} \min \quad x_1 + 2x_2, \\ \text{subject to} \begin{cases} 2x_1 + x_2 - x_3 = 1, \\ -3x_1 + x_2 + x_4 = 4, \\ x_1, x_2, x_3, x_4 \geqslant 0. \end{cases} \end{cases}$$

In the current case, $m = 2$, $n = 4$, and

$$A = \begin{pmatrix} 2 & 1 & -1 & 0 \\ -3 & 1 & 0 & 1 \end{pmatrix}, \quad b = \begin{pmatrix} 1 \\ 4 \end{pmatrix}, \quad c^\top = (1, 2, 0, 0).$$

To find extreme directions, we solve

$$\begin{cases} 2d_1 + d_2 - d_3 = 0, \\ -3d_1 + d_2 + d_4 = 0. \end{cases}$$

Thus,

$$d_3 = 2d_1 + d_2, \qquad d_4 = 3d_1 - d_2.$$

Then

$$d = \begin{pmatrix} d_1 \\ d_2 \\ 2d_1 + d_2 \\ 3d_1 - d_2 \end{pmatrix}.$$

Hence, such a d is an extreme direction if and only if

$$0 \leqslant d_2 \leqslant 3d_1.$$

Thus, the feasible set F of the problem admits non-zero extreme directions. Note that for any $d = (d_1, d_2, d_3, d_4)^\top \geqslant 0$, we have

$$c^\top d = d_1 + 2d_2 \geqslant 0.$$

By Theorem 2.15, the LP problem is bounded from below, i.e.,

$$\inf_{x \in F} c^\top x > -\infty.$$

Let us find the feasible set F of the problem. To this end, we solve the following:

$$\begin{cases} 2x_1 + x_2 - x_3 = 1, \\ -3x_1 + x_2 + x_4 = 4. \end{cases}$$

Clearly,
$$x_3 = 2x_1 + x_2 - 1, \qquad x_4 = 4 + 3x_1 - x_2.$$
Hence,
$$F = \left\{ \begin{pmatrix} x_1 \\ x_2 \\ 2x_1 + x_2 - 1 \\ 4 + 3x_1 - x_2 \end{pmatrix} \mid x_1, x_2 \geqslant 0, \ 2x_1 + x_2 \geqslant 1, \ 4 + 3x_1 - x_2 \geqslant 0 \right\}.$$
Since $m = 2$, any extreme point x can have at most 2 non-zero entries. Now, let $x = (x_1, x_2, x_3, x_4)^\top \in \mathcal{E}(F)$.

Case 1. If $x_1, x_2 > 0$, then we must have
$$\begin{cases} 2x_1 + x_2 - 1 = 0, \\ 4 + 3x_1 - x_2 = 0. \end{cases}$$
This will lead to $5x_1 + 3 = 0$. Thus, such a point is not feasible.

Case 2. If $x_1 = x_2 = 0$, then the corresponding point is not feasible.

Case 3. If $x_1 = 0$ and $x_2 > 0$, then
$$x = \begin{pmatrix} 0 \\ x_2 \\ x_2 - 1 \\ 4 - x_2 \end{pmatrix} \geqslant 0.$$
Thus,
$$1 \leqslant x_2 \leqslant 4.$$
Hence, we have extreme points (making use of Theorem 2.6, checking the linear independence of the column vectors corresponding to the positive components of the found vectors)
$$x^* = (0, 1, 0, 3)^\top, \qquad x^{**} = (0, 4, 3, 0)^\top.$$
Case 4. If $x_1 > 0, x_2 = 0$, then
$$x = \begin{pmatrix} x_1 \\ 0 \\ 2x_1 - 1 \\ 4 + 3x_1 \end{pmatrix} \geqslant 0.$$
Since $4 + 3x_1 > 0$, we must have $x_2 = \frac{1}{2}$. Thus, the extreme point is (making use of Theorem 2.6 again)
$$x^{***} = \left(\frac{1}{2}, 0, 0, \frac{11}{2} \right)^\top.$$
We now evaluate
$$c^\top x^* = 2, \quad c^\top x^{**} = 8, \quad c^\top x^{***} = \frac{1}{2}.$$
Hence, the minimum is attained at x^{***} with the minimum $\frac{1}{2}$.

Exercises

1. For given matrix A and vector b, determine the feasible set $F = \{x \in \mathbb{R}^n \mid Ax = b, \ x \geqslant 0\}$, and find the set of extreme points $\mathcal{E}(F)$ of F:

(i) $A = \begin{pmatrix} -2 & 0 & 3 \\ 1 & 2 & 4 \end{pmatrix}$, $b = \begin{pmatrix} -1 \\ 1 \end{pmatrix}$;

(ii) $A = \begin{pmatrix} 2 & 1 & -2 & 5 \\ 0 & 1 & 4 & -1 \end{pmatrix}$, $b = \begin{pmatrix} 0 \\ 4 \end{pmatrix}$;

2. Show that any polyhedron in $\mathbb{R}^n$ is convex and closed.

3. If $P_1, P_2, \cdots, P_k$ are polyhedrons in $\mathbb{R}^n$. Show that $\bigcap_{i=1}^{k} P_i$ is also a polyhedron in $\mathbb{R}^n$.

4. Show that the disk $B_1(0) = \{(x, y) \in \mathbb{R}^2 \mid x^2 + y^2 \leqslant 1\}$ can be written as the intersection of a sequence of polyhedrons.

5. Solve the following LP problem:
$$\begin{cases} \max \quad 25x_1 + 30x_2, \\ \text{subject to} \quad \begin{cases} 20x_1 + 30x_2 \leqslant 690, \\ 5x_1 + 4x_2 \leqslant 120, \\ x_1, x_2 \geqslant 0. \end{cases} \end{cases}$$

6. For given (A, b, c), using Theorem 5.2.8 to solve corresponding LP problem:

(i) $A = \begin{pmatrix} 0 & 1 & 3 \\ -1 & 0 & 6 \end{pmatrix}$, $b = \begin{pmatrix} 1 \\ 1 \end{pmatrix}$, $c = \begin{pmatrix} 1 \\ 2 \\ 3 \end{pmatrix}$.

(ii) $A = \begin{pmatrix} -1 & 1 & 1 & 0 \\ 3 & 0 & 1 & 1 \\ -2 & 1 & 0 & 1 \end{pmatrix}$, $b = \begin{pmatrix} 1 \\ 3 \\ 2 \end{pmatrix}$, $c = \begin{pmatrix} 1 \\ 0 \\ 1 \\ 2 \end{pmatrix}$.

7. Let A be given. Find all extreme directions for the feasible set F determined by A and some vector b:

(i) $A = \begin{pmatrix} 3 & 1 & 0 \\ 0 & 1 & -3 \end{pmatrix}$; (ii) $A = \begin{pmatrix} -1 & 2 & 0 & 3 \\ 1 & -2 & -6 & 0 \end{pmatrix}$.

8*. Let
$$F = \{(x, y) \in \mathbb{R}^2 \mid x + y \leqslant 5, \quad x - 2y \geqslant 1, \quad y \geqslant -1\}.$$
Find $\mathcal{E}(F)$. Then find maximum and minimum of $f(x) = x + 2y$, over F.

3 The Simplex Method

3.1 *General consideration*

From the fundamental theorem of linear programming, we see that to solve standard LP (1.1) which is bounded below, it suffices to find the set $\mathcal{E}(F)$ of all extreme points of the feasible set F. Then evaluate the objective function at these extreme points and make comparisons to get the minimum. Since we have n decision variables and m equality constraints, in general, we might have (by Theorem 2.6)

$$\binom{n}{m} = \frac{n!}{m!(n-m)!}$$

extreme points of F. The above number could be very large, for example, for $n = 20$ and $m = 10$, we have

$$\binom{20}{10} = \frac{20!}{(10!)^2} = 184,756$$

which is apparently too big to calculate. Therefore, some more effective methods are desirable.

In this section, we are going to present a method which gives a practically feasible search towards an optimal solution of LP.

For LP (1.1), due to the full rank condition (1.2) for A, after relabeling the components of the decision variable x and accordingly exchanging the columns of A (as well as the components of c), we may assume

$$A = (B, N), \qquad B \in \mathbb{R}^{m \times m}, \quad \det B \neq 0, \quad N \in \mathbb{R}^{m \times (n-m)}, \qquad (3.1)$$

and

$$x = \begin{pmatrix} x_B \\ x_N \end{pmatrix}, \qquad c = \begin{pmatrix} c_B \\ c_N \end{pmatrix}. \qquad (3.2)$$

We call B a *basis matrix* and N a *non-basis matrix*. Components in x_B are called *basic variables* and components in x_N are called *non-basic variables*. We refer to (3.1)– (3.2) as a *basic decomposition*. Note that by respectively exchanging the columns of B and/or N, we obtain different basic decompositions. However, such kind of changes do not change the set of basic/non-basic variables. Hence, we regard these basic decompositions *essentially the same*. Any two basic decompositions are said to be *essentially different* if they have different sets of basic variables. Since we assume $n > m$, there could be essentially different basic decompositions of LP (1.1). Now, under the basic decomposition (3.1)–(3.2), the equality constraint in (1.1) becomes

$$Bx_B + Nx_N = (B, N)\begin{pmatrix} x_B \\ x_N \end{pmatrix} = Ax = b. \tag{3.3}$$

Since B is invertible, we obtain

$$x_B = B^{-1}(b - Nx_N). \tag{3.4}$$

Hence, for any choice $x_N \in \mathbb{R}^{n-m}$, by defining x_B through (3.4), we obtain a general solution of the equality constraint in (1.1) as follows:

$$x = \begin{pmatrix} x_B \\ x_N \end{pmatrix} = \begin{pmatrix} B^{-1}(b - Nx_N) \\ x_N \end{pmatrix}.$$

In particular, by taking $x_N = 0$ in the above, we have that

$$x^* \triangleq \begin{pmatrix} x_B^* \\ x_N^* \end{pmatrix} \equiv \begin{pmatrix} B^{-1}b \\ 0 \end{pmatrix} \tag{3.5}$$

is a solution of the linear equation in (1.1). Such an x^* is called a *basic solution* of (1.1), which might not be feasible since $x_B^* \geqslant 0$ is not guaranteed. Now, if, in addition, the following holds:

$$x_B^* = B^{-1}b \geqslant 0,$$

then x^* defined by (3.5) is called a *basic feasible solution* of (1.1). Note that the columns of A corresponding to the positive components of x^*, which must be some (or all) components of x_B^*, are some (or all) columns of B. They are linearly independent, since B itself is invertible. Hence, by Theorem 2.6, x^* must be an extreme point of F. Further, if

$$x_B^* \equiv B^{-1}b > 0, \tag{3.6}$$

then x^* is said to be *non-degenerate*. Otherwise, it is said to be *degenerate*. If all the basic feasible solutions of a standard form LP are non-degenerate, then the LP is said to be . Note that the non-degeneracy condition of an LP only depends on A and b. Thus it is intrinsic.

Example 3.1. Let

$$A = \begin{pmatrix} 2 & 1 & 3 & 0 \\ 0 & 1 & 1 & 1 \end{pmatrix}, \quad b = \begin{pmatrix} 1 \\ 2 \end{pmatrix}.$$

Let

$$B = \begin{pmatrix} 2 & 1 \\ 0 & 1 \end{pmatrix}, \quad N = \begin{pmatrix} 3 & 0 \\ 1 & 1 \end{pmatrix}, \quad x_B = \begin{pmatrix} x_1 \\ x_2 \end{pmatrix}, \quad x_N = \begin{pmatrix} x_3 \\ x_4 \end{pmatrix}.$$

Then

$$x_B = \begin{pmatrix} x_1 \\ x_2 \end{pmatrix} = B^{-1}\left[b - Nx_N\right] = \begin{pmatrix} \frac{1}{2} & -\frac{1}{2} \\ 0 & 1 \end{pmatrix}\left[\begin{pmatrix} 1 \\ 2 \end{pmatrix} - \begin{pmatrix} 3 & 0 \\ 1 & 1 \end{pmatrix}\begin{pmatrix} x_3 \\ x_4 \end{pmatrix}\right].$$

Hence,

$$x_B^* = \begin{pmatrix} \frac{1}{2} & -\frac{1}{2} \\ 0 & 1 \end{pmatrix}\begin{pmatrix} 1 \\ 2 \end{pmatrix} = \begin{pmatrix} -\frac{1}{2} \\ 2 \end{pmatrix}, \qquad x_N^* = 0.$$

Clearly, $(x_1, x_2, x_3, x_4) = (-\frac{1}{2}, 2, 0, 0)$ is a basic solution, but it is not feasible.

Now, we order the components of x as (x_2, x_4, x_1, x_3), then correspondingly,

$$\widetilde{A} = \begin{pmatrix} 1 & 0 & 2 & 3 \\ 1 & 1 & 0 & 1 \end{pmatrix}, \quad \widetilde{x}_B = \begin{pmatrix} x_2 \\ x_4 \end{pmatrix}, \quad \widetilde{x}_N = \begin{pmatrix} x_1 \\ x_3 \end{pmatrix},$$

and

$$\widetilde{B} = \begin{pmatrix} 1 & 0 \\ 1 & 1 \end{pmatrix}, \quad \widetilde{N} = \begin{pmatrix} 2 & 3 \\ 0 & 1 \end{pmatrix}.$$

Hence,

$$\widetilde{x}_B = \begin{pmatrix} x_2 \\ x_4 \end{pmatrix} = \widetilde{B}^{-1}\left[b - \widetilde{N}\widetilde{x}_N\right] = \begin{pmatrix} 1 & 0 \\ -1 & 1 \end{pmatrix}\left[\begin{pmatrix} 1 \\ 2 \end{pmatrix} - \begin{pmatrix} 2 & 3 \\ 0 & 1 \end{pmatrix}\begin{pmatrix} x_1 \\ x_3 \end{pmatrix}\right].$$

Then by taking $\widetilde{x}_N = 0$, we have

$$\widetilde{x}_B^* = \begin{pmatrix} x_2 \\ x_4 \end{pmatrix} = \begin{pmatrix} 1 & 0 \\ -1 & 1 \end{pmatrix}\begin{pmatrix} 1 \\ 2 \end{pmatrix} = \begin{pmatrix} 1 \\ 1 \end{pmatrix} > 0.$$

Hence, $(x_1, x_2, x_3, x_4) = (0, 1, 0, 1)$ is a basic solution which is feasible.

Now, we return to the general problem. Under the basic decomposition of form (3.1)–(3.2), we have the following:

$$c^\top x = (c_B^\top, c_N^\top)\begin{pmatrix} x_B \\ x_N \end{pmatrix} = c_B^\top x_B + c_N^\top x_N$$

$$= c_B^\top B^{-1}(b - Nx_N) + c_N^\top x_N = c_B^\top B^{-1}b + (c_N^\top - c_B^\top B^{-1}N)x_N \qquad (3.7)$$

$$= c_B^\top x_B^* + \begin{pmatrix} 0 \\ c_N - (B^{-1}N)^\top c_B \end{pmatrix}^\top \begin{pmatrix} x_B \\ x_N \end{pmatrix} = c^\top x^* + r^\top x,$$

where

$$r \equiv \begin{pmatrix} r_1 \\ r_2 \\ \vdots \\ r_n \end{pmatrix} = \begin{pmatrix} 0 \\ c_N - (B^{-1}N)^\top c_B \end{pmatrix} \equiv \begin{pmatrix} 0 \\ r_N \end{pmatrix},$$

which is called the *reduced cost vector* (corresponding to the basic decomposition (3.1)–(3.2)). We have the following result.

Proposition 3.2. *Suppose the feasible set F of standard form LP (1.1) is non-empty. Suppose*

$$\inf_{x \in F} c^\top x > -\infty.$$

Then there exists a basic decomposition of form (3.1)–(3.2), so that x^* defined by (3.5) is an optimal solution of LP (1.1). Moreover, under any basic decomposition of form (3.1)–(3.2), if x^* given by (3.5) is feasible, and

$$r_N \equiv c_N - (B^{-1}N)^\top c_B \geq 0, \tag{3.8}$$

then x^* is optimal. Conversely, if x^* defined by (3.5) is optimal and is non-degenerate, i.e., (3.6) holds, then (3.8) holds.

Proof. By Theorem 2.16, we know that there exists an extreme point x^* of F at which the objective function achieves its minimum over F. Without loss of generality, we assume that

$$x^* = \begin{pmatrix} x_0^* \\ 0 \end{pmatrix},$$

with $x_0^* \in \mathbb{R}_{++}^k$, $1 \leq k \leq m$. Correspondingly,

$$A = (\widetilde{B}, \widetilde{N}), \qquad \widetilde{B} \in \mathbb{R}^{m \times k}, \quad \widetilde{N} \in \mathbb{R}^{m \times (n-k)}.$$

By Theorem 2.6, we see that the columns of A corresponding to the positive components of x^* are linearly independent, i.e.,

$$\text{rank}(\widetilde{B}) = k.$$

Now, by the condition $\text{rank}(A) = m$, we can select $m - k$ columns from $\widetilde{N}$ (if $k < m$, otherwise it is not necessary) so that these columns together the columns of $\widetilde{B}$ form an invertible $(m \times m)$ matrix. Hence, we can find a basic decomposition of form (3.1)–(3.2) so that (3.5) holds.

Next, under any basic decomposition of form (3.1)–(3.2), for any feasible solution $x = \begin{pmatrix} x_B \\ x_N \end{pmatrix}$, it follows from (3.7) that (note condition (3.8))

$$c^\top x = c_B B^{-1} b + (c_N^\top - c_B^\top B^{-1} N) x_N = c^\top x^* + [c_N - (B^{-1}N)^\top c_B]^\top x_N$$
$$= c^\top x^* + r_N^\top x_N \geq c^\top x^*, \tag{3.9}$$

proving the optimality of x^*.

Finally, if (3.6) holds, then

$$x_B = B^{-1}b - B^{-1}Nx_N = x_B^* - B^{-1}Nx_N > 0,$$

provided $x_N \geqslant 0$ is small enough. Hence, the optimality of x^*, which is exhibited in (3.9) implies

$$r_N^\top x_N \geqslant 0,$$

for any $x_N \geqslant 0$ small enough. Hence, (3.8) follows.   □

Let us look at the following example.

Example 3.3. Let

$$A = \begin{pmatrix} 1 & 0 & 1 & 1 \\ 1 & 1 & 2 & 1 \end{pmatrix}, \quad b = \begin{pmatrix} 1 \\ 2 \end{pmatrix}, \quad c^\top = (1, 1, \alpha, \beta),$$

where $\alpha, \beta \in \mathbb{R}$. Let

$$B = \begin{pmatrix} 1 & 0 \\ 1 & 1 \end{pmatrix}, \quad N = \begin{pmatrix} 1 & 1 \\ 2 & 1 \end{pmatrix}, \quad x_B = \begin{pmatrix} x_1 \\ x_2 \end{pmatrix}, \quad x_N = \begin{pmatrix} x_3 \\ x_4 \end{pmatrix}.$$

Then

$$x_B = \begin{pmatrix} x_1 \\ x_2 \end{pmatrix} = B^{-1}\big[b - Nx_N\big] = \begin{pmatrix} 1 & 0 \\ -1 & 1 \end{pmatrix} \left[\begin{pmatrix} 1 \\ 2 \end{pmatrix} - \begin{pmatrix} 1 & 1 \\ 2 & 1 \end{pmatrix} \begin{pmatrix} x_3 \\ x_4 \end{pmatrix} \right].$$

Consequently,

$$x_B^* = \begin{pmatrix} 1 & 0 \\ -1 & 1 \end{pmatrix} \begin{pmatrix} 1 \\ 2 \end{pmatrix} = \begin{pmatrix} 1 \\ 1 \end{pmatrix}.$$

Thus, x^* is a basic feasible solution. Now,

$$r_N = c_N - (B^{-1}N)^\top c_B = \begin{pmatrix} \alpha \\ \beta \end{pmatrix} - \left[\begin{pmatrix} 1 & 0 \\ -1 & 1 \end{pmatrix} \begin{pmatrix} 1 & 1 \\ 2 & 1 \end{pmatrix} \right]^\top \begin{pmatrix} 1 \\ 1 \end{pmatrix} = \begin{pmatrix} \alpha - 2 \\ \beta - 1 \end{pmatrix}.$$

Thus, if $\alpha \geqslant 2$ and $\beta \geqslant 1$, then $r_N \geqslant 0$, and x^* is optimal. Otherwise, $r_N \geqslant 0$ fails, $x^* = (1, 1, 0, 0)^\top$ is not optimal.

We point out that when (3.6) fails, we do not have the implication of (3.8) from the optimality of x^*, in general. Here is an example.

Example 3.4. Consider standard form LP:

$$\begin{cases} \min & 2x_1 + x_2, \\ \text{subject to} & x_1 + x_2 = 0, \qquad x_1, x_2 \geqslant 0. \end{cases}$$

Clearly, $x_1 = x_2 = 0$ is the optimal solution which is degenerate. We now let x_1 be the basic variable and x_2 be the non-basic variable. Then

$$B = N = 1, \qquad c_B = 2, \qquad c_N = 1.$$

Hence,

$$c_N - (B^{-1}N)^\top c_B = 1 - 2 = -1 < 0.$$

Thus, (3.8) fails in this case.

3.2 Phase II

Now, suppose we have basic decomposition (3.1)–(3.2). We would like to investigate what we should do if a feasible basic solution x^* is found and $r_N \geqslant 0$ fails. To begin with, let $\Lambda_B = \{1, \cdots, m\}$ and $\Lambda_N \triangleq \{m+1, \cdots, n\}$. Clearly, condition (3.8) can be rewritten as

$$r_j \geqslant 0, \qquad \forall j \in \Lambda_N. \tag{3.10}$$

We have the following result.

Proposition 3.5. *Let A be decomposed as (3.1) with $N = (a_{m+1}, \cdots, a_n)$. Let x^* defined by (3.5) be feasible and non-degenerate. Suppose for some $q \in \Lambda_N$,*

$$r_q < 0. \tag{3.11}$$

Then the following conclusions hold.

(i) *If it holds*

$$B^{-1} a_q \leqslant 0, \tag{3.12}$$

then the LP is unbounded below, i.e.,

$$\inf\{c^\top x \mid x \in F\} = -\infty. \tag{3.13}$$

(ii) *If LP is bounded below, then there exists an $x^{**} \in \mathcal{E}(F)$ such that*

$$c^\top x^{**} < c^\top x^*. \tag{3.14}$$

Proof. According to Proposition 3.2, since x^* non-degenerate, (3.11) implies that x^* is not optimal. We define

$$\xi^q \equiv \begin{pmatrix} \xi_1^q \\ \vdots \\ \xi_n^q \end{pmatrix} \triangleq \begin{pmatrix} -B^{-1} a_q \\ \varepsilon_q \end{pmatrix} \in \mathbb{R}^n. \tag{3.15}$$

Hereafter, $\varepsilon_j \in \mathbb{R}^{n-m}$ is the vector with the components labeled by Λ_N such that the j-th entry is 1 and all others are 0 ($j \in \Lambda_N$). Clearly,

$$A\xi^q = (B, N) \begin{pmatrix} -B^{-1} a_q \\ \varepsilon_q \end{pmatrix} = -a_q + N\varepsilon_q = -a_q + a_q = 0.$$

Thus, for any $\lambda > 0$, the following:

$$x^{**} \triangleq x^* + \lambda \xi^q = \begin{pmatrix} B^{-1} b \\ 0 \end{pmatrix} + \lambda \begin{pmatrix} -B^{-1} a_q \\ \varepsilon_q \end{pmatrix} = \begin{pmatrix} B^{-1} b - \lambda B^{-1} a_q \\ \lambda \varepsilon_q \end{pmatrix} \tag{3.16}$$

satisfies the equality constraint in (1.1). Further, in order x^{**} defined by (3.16) to be feasible, we need

$$B^{-1}b - \lambda B^{-1}a_q \geqslant 0. \tag{3.17}$$

Hence, in the case that (3.12) holds, (3.17) holds for any $\lambda > 0$. Consequently, due to (3.11), we obtain

$$
\begin{aligned}
c^\top x^{**} &= c^\top (x^* + \lambda \xi^q) = (c_B^\top, c_N^\top) \begin{pmatrix} B^{-1}b - \lambda B^{-1}a_q \\ \lambda \varepsilon_q \end{pmatrix} \\
&= c_B^\top (B^{-1}b - \lambda B^{-1}a_q) + \lambda c_N^\top \varepsilon_q = c_B^\top x_B^* + \lambda (c_N^\top \varepsilon_q - c_B^\top B^{-1}a_q) \\
&= c^\top x^* + \lambda (c_N^\top - c_B^\top B^{-1}N)\varepsilon_q = c^\top x^* + \lambda r_q \to -\infty.
\end{aligned} \tag{3.18}
$$

This proves (i).

In the case that the LP is bounded from below, the case (i) will not happen. Pick a $q \in \Lambda_N$ such that (3.11) holds. Since (3.12) fails, for some $i \in \Lambda_B$, $-\xi_i^q \equiv (B^{-1}a_q)_i > 0$, where $(B^{-1}a_q)_i$ is the i-th component of $B^{-1}a_q$. Then we can find a $p \in \Lambda_B$ such that

$$
\begin{aligned}
\lambda &\triangleq \min_{i \in \Lambda_B} \left\{ \frac{x_i^*}{-\xi_i^q} \ \Big| \ \xi_i^q < 0 \right\} \equiv \min_{i \in \Lambda_B} \left\{ \frac{x_i^*}{(B^{-1}a_q)_i} \ \Big| \ (B^{-1}a_q)_i > 0 \right\} \\
&= \frac{x_p^*}{(B^{-1}a_q)_p} \equiv \frac{x_p^*}{-\xi_p^q}.
\end{aligned} \tag{3.19}
$$

Since x^* is non-degenerate, $x_p^* > 0$. Recall $e_k \in \mathbb{R}^n$, the vector with the k-th entry being 1 and all others being zero. By taking the *step size* λ as (3.19), called a *minimum ratio*, and defining x^{**} by (3.16), we have

$$x_k^{**} = e_k^\top x^{**} = e_k^\top (x^* + \lambda \xi^q) = x_k^* + \lambda \xi_k^q \geqslant 0, \quad 1 \leqslant k \leqslant n,$$

and

$$x_p^{**} = e_p^\top x^{**} = e_p^\top (x^* + \lambda \xi^q) = x_p^* + \lambda \xi_p^q = 0. \tag{3.20}$$

From the above, we see that $x^{**} \in F$. Then by (3.18) and (3.11), we obtain (3.14). Finally, we want to show that $x^{**} \in \mathcal{E}(F)$. To this end, we note that

$$x_k^{**} = x_k^* = 0, \quad k \in [\Lambda_N \setminus \{q\}] \bigcup \{p\} \triangleq \Lambda_{\bar{N}}.$$

Let $\Lambda_{\bar{B}} = [\Lambda_B \setminus \{p\}] \bigcup \{q\}$. We claim that $\{a_k, k \in \Lambda_{\bar{B}}\}$ are linearly independent. In fact, we have

$$\sum_{k \in \Lambda_B,\, k \neq p} x_k^* a_k + x_p^* a_p = Ax^* = b = Ax^{**} = \sum_{k \in \Lambda_B,\, k \neq p} x_k^{**} a_k + x_q^{**} a_q.$$

Thus,

$$a_p = \sum_{k \in \Lambda_B, k \neq p} \frac{x_k^{**} - x_k^*}{x_p^*} a_k + \frac{x_q^{**}}{x_p^*} a_q.$$

Since $\{a_k \mid k \in \Lambda_B\}$ are linearly independent, we must have the linear independence of $\{a_k \mid k \in \Lambda_{\bar{B}}\}$. Thus, by Theorem 2.6, we see that $x^{**} \in \mathcal{E}(F)$. □

We see that when LP is bounded below, $x^{**} \in F$ is a better feasible solution than x^*. Therefore, ξ^q is called an *improvement direction*. In the process of moving from x^* to x^{**}, we see that variable x_p leaves the basis and variable x_q enters the basis. The index $q \in \Lambda_N$ for x_q entering the basis is determined by (3.11). Naturally, we should choose a $q \in \Lambda_N$ so that r_q is the most negative among all r_q's (because it gives the biggest reduction of the objective function value). If there are more than one such q, we may choose one with the smallest index. We refer to this as the *largest reduction cost (with smallest index) rule*. The index $p \in \Lambda_B$ for x_p leaving the basis is determined by the minimum ratio. If there are more than one such p, we may choose one with the smallest index. We refer to this as the *minimum ratio with smallest index rule*.

After moving from x^* to x^{**}, if the reduced cost vector corresponding to x^{**} is non-negative, then x^{**} is an optimal solution. Otherwise, we can repeat the above procedure. Since each time, we have reduced the objective function value (see below for more considerations on the degenerate case), and the set $\mathcal{E}(F)$ of extreme points of F is finite, after finitely many steps, we expect to end up with an optimal solution to the LP.

The above is called the *Phase II simplex method*, or *Phase II problem* which can be applied if an initial basic feasible solution is known (or can be easily found). Note that one always transforms an LP problem into a corresponding standard form before applying the simplex method. We now present an example to illustrate the (Phase II) simplex method.

Example 3.6. Consider the following LP in its standard form:

$$\begin{cases} \min & -2x_1 - 3x_2, \\ \text{subject to} & \begin{cases} -x_1 + x_2 + x_3 = 5, \\ x_1 + 3x_2 + x_4 = 35, \\ x_1 + x_5 = 20, \\ x_1, x_2, x_3, x_4, x_5 \geqslant 0. \end{cases} \end{cases}$$

For this problem, we have

$$A = \begin{pmatrix} -1 & 1 & 1 & 0 & 0 \\ 1 & 3 & 0 & 1 & 0 \\ 1 & 0 & 0 & 0 & 1 \end{pmatrix}, \quad b = \begin{pmatrix} 5 \\ 35 \\ 20 \end{pmatrix}, \quad c = \begin{pmatrix} -2 \\ -3 \\ 0 \\ 0 \\ 0 \end{pmatrix}.$$

It is obvious that $x^* = (0,0,5,35,20)^\top$ is a basic feasible solution. Therefore, we first choose x_3, x_4, x_5 as basic variables, and x_1, x_2 as non-basic variables. Then we may choose the following basic decomposition:

$$\begin{cases} B = \begin{pmatrix} 1 & 0 & 0 \\ 0 & 1 & 0 \\ 0 & 0 & 1 \end{pmatrix}, \quad N = \begin{pmatrix} -1 & 1 \\ 1 & 3 \\ 1 & 0 \end{pmatrix}, \quad c_B = \begin{pmatrix} 0 \\ 0 \\ 0 \end{pmatrix}, \quad c_N = \begin{pmatrix} -2 \\ -3 \end{pmatrix}, \\[2mm] x = \begin{pmatrix} x_B \\ x_N \end{pmatrix}, \quad x_B \triangleq \begin{pmatrix} x_3 \\ x_4 \\ x_5 \end{pmatrix}, \quad x_N \triangleq \begin{pmatrix} x_1 \\ x_2 \end{pmatrix}, \quad \Lambda_B = \{3,4,5\}, \quad \Lambda_N = \{1,2\}. \end{cases}$$

Under the above basic decomposition, we have the current feasible basic solution:

$$x^{(1)} \equiv \begin{pmatrix} x_3^{(1)} \\ x_4^{(1)} \\ x_5^{(1)} \\ x_1^{(1)} \\ x_2^{(1)} \end{pmatrix} = \begin{pmatrix} x_B^{(1)} \\ 0 \end{pmatrix} = \begin{pmatrix} B^{-1}b \\ 0 \end{pmatrix} = \begin{pmatrix} 5 \\ 35 \\ 20 \\ 0 \\ 0 \end{pmatrix},$$

with the objective function value

$$[c^{(1)}]^\top x^{(1)} = (0\ 0\ 0\ -2\ -3) \begin{pmatrix} 5 \\ 35 \\ 20 \\ 0 \\ 0 \end{pmatrix} = 0. \tag{3.21}$$

Note that $c^{(1)}$ is obtained by relabeling according to that of x. Since

$$r_N \equiv c_N - (B^{-1}N)^\top c_B = \begin{pmatrix} -2 \\ -3 \end{pmatrix},$$

the condition $r_N \geqslant 0$ fails. Hence, the current solution $x^{(1)}$ is not optimal. We pick $q = 2$, since $-3 < -2$, the unit cost reduction for x_2 is larger than that of x_1 (thus, x_2 will enter the basis). Then noting $a_2^\top = (1, 3, 0)$, we have

$$\xi^2 = \begin{pmatrix} -B^{-1}a_2 \\ \varepsilon_2 \end{pmatrix} = \begin{pmatrix} -1 \\ -3 \\ 0 \\ 0 \\ 1 \end{pmatrix}.$$

Now the minimum ratio is given by

$$\lambda = \min\left\{ \frac{x_3^{(1)}}{-\xi_3^2}, \frac{x_4^{(1)}}{-\xi_4^2} \right\} = \min\left\{ \frac{5}{1}, \frac{35}{3} \right\} = 5 \equiv \frac{x_3^{(1)}}{-\xi_3^2}.$$

Thus, x_3 will leave the basis. The next basic feasible solution is

$$x^{(2)} \equiv \begin{pmatrix} x_3^{(2)} \\ x_4^{(2)} \\ x_5^{(2)} \\ x_1^{(2)} \\ x_2^{(2)} \end{pmatrix} = x^{(1)} + \lambda\xi^2 = \begin{pmatrix} 5 \\ 35 \\ 20 \\ 0 \\ 0 \end{pmatrix} + 5 \begin{pmatrix} -1 \\ -3 \\ 0 \\ 0 \\ 1 \end{pmatrix} = \begin{pmatrix} 0 \\ 20 \\ 20 \\ 0 \\ 5 \end{pmatrix} \geqslant 0.$$

The new basic variables are x_2, x_4 and x_5 (x_3 leaves), and new non-basic variables are x_1 and x_3. Now, we take the basic decomposition

$$\begin{cases} B = \begin{pmatrix} 1 & 0 & 0 \\ 3 & 1 & 0 \\ 0 & 0 & 1 \end{pmatrix}, \ N = \begin{pmatrix} -1 & 1 \\ 1 & 0 \\ 1 & 0 \end{pmatrix}, \ c_B = \begin{pmatrix} -3 \\ 0 \\ 0 \end{pmatrix}, \ c_N = \begin{pmatrix} -2 \\ 0 \end{pmatrix}, \\ x = \begin{pmatrix} x_B \\ x_N \end{pmatrix}, \ x_B \triangleq \begin{pmatrix} x_2 \\ x_4 \\ x_5 \end{pmatrix}, \ x_N \triangleq \begin{pmatrix} x_1 \\ x_3 \end{pmatrix}, \ \Lambda_B = \{2, 4, 5\}, \ \Lambda_N = \{1, 3\}. \end{cases}$$

Then $B^{-1} = \begin{pmatrix} 1 & 0 & 0 \\ -3 & 1 & 0 \\ 0 & 0 & 1 \end{pmatrix}$, and under the new basic decomposition, we can rewrite $x^{(2)}$ as

$$x^{(2)} \equiv \begin{pmatrix} x_B^{(2)} \\ x_N^{(2)} \end{pmatrix} \equiv \begin{pmatrix} x_2^{(2)} \\ x_4^{(2)} \\ x_5^{(2)} \\ x_1^{(2)} \\ x_3^{(2)} \end{pmatrix} = \begin{pmatrix} 5 \\ 20 \\ 20 \\ 0 \\ 0 \end{pmatrix} = \begin{pmatrix} B^{-1}b \\ 0 \end{pmatrix},$$

with the objective function value

$$[c^{(2)}]^\top x^{(2)} = (-3\ 0\ 0\ -2\ 0) \begin{pmatrix} 5 \\ 20 \\ 20 \\ 0 \\ 0 \end{pmatrix} = -15, \tag{3.22}$$

which is smaller than $[c^{(1)}]^\top x^{(1)}$ (see (3.21)). We compute

$$r_N \equiv c_N - (B^{-1}N)^\top c_B = \begin{pmatrix} -2 \\ 0 \end{pmatrix} - \begin{pmatrix} -1 & 4 & 1 \\ 1 & -3 & 0 \end{pmatrix} \begin{pmatrix} -3 \\ 0 \\ 0 \end{pmatrix} = \begin{pmatrix} -5 \\ 3 \end{pmatrix}.$$

Since $r_N \geqslant 0$ fails again, $x^{(2)}$ is not optimal yet. We pick $q = 1$ since this is the only q with $r_q < 0$ (thus x_1 will enter the basis). Then

$$\xi^1 = \begin{pmatrix} -B^{-1}a_1 \\ \varepsilon_1 \end{pmatrix} = \begin{pmatrix} 1 \\ -4 \\ -1 \\ 1 \\ 0 \end{pmatrix}.$$

Thus, the minimum ratio is given by

$$\lambda = \min\left\{\frac{x_4^{(2)}}{-\xi_4^1}, \frac{x_5^{(2)}}{-\xi_5^1}\right\} = \min\left\{\frac{20}{4}, \frac{20}{1}\right\} = 5 \equiv \frac{x_4^{(2)}}{-\xi_4^1},$$

and the next basic feasible solution is

$$x^{(3)} \equiv \begin{pmatrix} x_2^{(3)} \\ x_4^{(3)} \\ x_5^{(3)} \\ x_1^{(3)} \\ x_3^{(3)} \end{pmatrix} = x^{(2)} + \lambda\xi^1 = \begin{pmatrix} 5 \\ 20 \\ 20 \\ 0 \\ 0 \end{pmatrix} + 5 \begin{pmatrix} 1 \\ -4 \\ -1 \\ 1 \\ 0 \end{pmatrix} = \begin{pmatrix} 10 \\ 0 \\ 15 \\ 5 \\ 0 \end{pmatrix} \geqslant 0.$$

The new basic variables are x_1, x_2, and x_5 (x_4 leaves), and the new non-basic variables are x_2 and x_4. Now, we take the basic decomposition

$$\begin{cases} B = \begin{pmatrix} -1 & 1 & 0 \\ 1 & 3 & 0 \\ 1 & 0 & 1 \end{pmatrix}, \ N = \begin{pmatrix} 1 & 0 \\ 0 & 1 \\ 0 & 0 \end{pmatrix}, \ c_B = \begin{pmatrix} -2 \\ -3 \\ 0 \end{pmatrix}, \ c_N = \begin{pmatrix} 0 \\ 0 \end{pmatrix}, \\[4mm] x = \begin{pmatrix} x_B \\ x_N \end{pmatrix}, \quad x_B \triangleq \begin{pmatrix} x_1 \\ x_2 \\ x_5 \end{pmatrix}, \quad x_N \triangleq \begin{pmatrix} x_3 \\ x_4 \end{pmatrix}, \quad \Lambda_B = \{1,2,5\}, \quad \Lambda_N = \{3,4\}. \end{cases}$$

Then $B^{-1} = \begin{pmatrix} -\frac{3}{4} & \frac{1}{4} & 0 \\ \frac{1}{4} & \frac{1}{4} & 0 \\ \frac{3}{4} & -\frac{1}{4} & 1 \end{pmatrix}$, and under the new basic decomposition, we can rewrite $x^{(3)}$ as

$$x^{(3)} \equiv \begin{pmatrix} x_B^{(3)} \\ x_N^{(3)} \end{pmatrix} \equiv \begin{pmatrix} x_1^{(3)} \\ x_2^{(3)} \\ x_5^{(3)} \\ x_3^{(3)} \\ x_4^{(2)} \end{pmatrix} = \begin{pmatrix} 5 \\ 10 \\ 15 \\ 0 \\ 0 \end{pmatrix} = \begin{pmatrix} B^{-1}b \\ 0 \end{pmatrix},$$

with the objective function value

$$[c^{(3)}]^\top x^{(3)} = (-2 \; -3 \; 0 \; 0 \; 0) \begin{pmatrix} 5 \\ 10 \\ 15 \\ 0 \\ 0 \end{pmatrix} = -40, \qquad (3.23)$$

which is smaller than $[c^{(2)}]^\top x^{(2)}$ (see (3.22)). We compute

$$r_N \equiv c_N - (B^{-1}N)^\top c_B = \begin{pmatrix} 0 \\ 0 \end{pmatrix} - \begin{pmatrix} -\frac{3}{4} & \frac{1}{4} & \frac{3}{4} \\ \frac{1}{4} & \frac{1}{4} & -\frac{1}{4} \end{pmatrix} \begin{pmatrix} -2 \\ -3 \\ 0 \end{pmatrix} = \begin{pmatrix} -\frac{3}{4} \\ \frac{5}{4} \end{pmatrix}.$$

Again, since $r_N \geq 0$ does not hold, $x^{(3)}$ is still not optimal. We now pick $q = 3$ since $r_3 < 0$ (thus x_3 will enter the basis). Then

$$\xi^3 = \begin{pmatrix} -B^{-1}a_3 \\ \varepsilon_3 \end{pmatrix} = \begin{pmatrix} \frac{3}{4} \\ -\frac{1}{4} \\ -\frac{3}{4} \\ 1 \\ 0 \end{pmatrix}.$$

Thus, the minimum ratio is given by

$$\lambda = \min\left\{ \frac{x_2^{(3)}}{-\xi_2^3}, \frac{x_5^{(3)}}{-\xi_5^3} \right\} = \min\left\{ \frac{10}{1/4}, \frac{15}{3/4} \right\} = 20 \equiv \frac{x_5^{(3)}}{-\xi_5^3},$$

and the next basic feasible solution is

$$x^{(4)} \equiv \begin{pmatrix} x_1^{(4)} \\ x_2^{(4)} \\ x_5^{(4)} \\ x_3^{(4)} \\ x_4^{(4)} \end{pmatrix} = x^{(3)} + \lambda\xi^3 = \begin{pmatrix} 5 \\ 10 \\ 15 \\ 0 \\ 0 \end{pmatrix} + 20 \begin{pmatrix} \frac{3}{4} \\ -\frac{1}{4} \\ -\frac{3}{4} \\ 1 \\ 0 \end{pmatrix} = \begin{pmatrix} 20 \\ 5 \\ 0 \\ 20 \\ 0 \end{pmatrix} \geq 0.$$

Hence, the new basic variables are x_1, x_2, and x_3 (x_5 leaves), and the new non-basic variables are x_4 and x_5. Now, we take the basic decomposition

$$
\begin{cases}
B = \begin{pmatrix} -1 & 1 & 1 \\ 1 & 3 & 0 \\ 1 & 0 & 0 \end{pmatrix}, \quad N = \begin{pmatrix} 0 & 0 \\ 1 & 0 \\ 0 & 1 \end{pmatrix}, \quad c_B = \begin{pmatrix} -2 \\ -3 \\ 0 \end{pmatrix}, \quad c_N = \begin{pmatrix} 0 \\ 0 \end{pmatrix}, \\[12pt]
x = \begin{pmatrix} x_B \\ x_N \end{pmatrix}, \quad x_B \triangleq \begin{pmatrix} x_1 \\ x_2 \\ x_3 \end{pmatrix}, \quad x_N \triangleq \begin{pmatrix} x_4 \\ x_5 \end{pmatrix}, \quad \Lambda_B = \{1,2,3\}, \quad \Lambda_N = \{4,5\}.
\end{cases}
$$

Then $B^{-1} = \begin{pmatrix} 0 & 0 & 1 \\ 0 & \frac{1}{3} & -\frac{1}{3} \\ 1 & -\frac{1}{3} & \frac{4}{3} \end{pmatrix}$ and under the new basic decomposition, one can

rewrite $x^{(4)}$ as

$$
x^{(4)} \equiv \begin{pmatrix} x_B^{(4)} \\ x_N^{(4)} \end{pmatrix} = \begin{pmatrix} x_1^{(4)} \\ x_2^{(4)} \\ x_3^{(4)} \\ x_4^{(4)} \\ x_5^{(4)} \end{pmatrix} = \begin{pmatrix} 20 \\ 5 \\ 20 \\ 0 \\ 0 \end{pmatrix},
$$

with the objective function value

$$
[c^{(4)}]^\top x^{(4)} = (-2 \ -3 \ 0 \ 0 \ 0) \begin{pmatrix} 20 \\ 5 \\ 20 \\ 0 \\ 0 \end{pmatrix} = -55, \tag{3.24}
$$

again, which is smaller that $[c^{(3)}]^\top x^{(3)}$ (see (3.23)). Now, we compute

$$
r_N \equiv c_N - (B^{-1}N)^\top c_B = \begin{pmatrix} 0 \\ 0 \end{pmatrix} - \begin{pmatrix} 0 & \frac{1}{3} & -\frac{1}{3} \\ 1 & -\frac{1}{3} & \frac{4}{3} \end{pmatrix} \begin{pmatrix} -2 \\ -3 \\ 0 \end{pmatrix} = \begin{pmatrix} 1 \\ 1 \end{pmatrix}.
$$

Since $r_N \geqslant 0$, solution $x^{(4)}$ is optimal. Therefore, the optimal solution is given by

$$
x_1 = 20, \quad x_2 = 5, \quad x_3 = 20, \quad x_4 = x_5 = 0,
$$

with the optimal objective function value

$$
f = -2x_1 - 3x_2 = -55.
$$

In the above example, at the second step, we return to the original system with a new basic decomposition, redo the same thing as the first step. But, it seems to be more reasonable to keep the reduced tableau from the first step (because the objective function value has already reached at a reduced value).

The above procedure seems to be a little too complicated. We now summarize the procedure using tableau:

(1) Suppose an initial basic feasible solution x^* is known. Let x^* have a representation (3.5), under a basic decomposition of form (3.1)–(3.2). We may also use the following tableau representation:

$$\begin{pmatrix} A & b \\ c^\top & f \end{pmatrix} = \begin{pmatrix} B & N & b \\ c_B^\top & c_N^\top & f \end{pmatrix},$$

which is understood as the following non-homogeneous linear equation:

$$\begin{pmatrix} B & N \\ c_B^\top & c_N^\top \end{pmatrix} \begin{pmatrix} x_B^* \\ 0 \end{pmatrix} = \begin{pmatrix} b \\ f \end{pmatrix}.$$

Since $x^* = \begin{pmatrix} x_B^* \\ 0 \end{pmatrix}$ and $c = \begin{pmatrix} c_B \\ c_N \end{pmatrix}$, the above can also be read as

$$Ax^* = b, \quad f = c^\top x^*.$$

Therefore, f is the value of the objective function corresponding the basic feasible solution x^*. Next, applying elementary row operations, we get

$$\begin{pmatrix} B^{-1} & 0 \\ -c_B^\top B^{-1} & 1 \end{pmatrix} \begin{pmatrix} B & N & b \\ c_B^\top & c_N^\top & f \end{pmatrix} = \begin{pmatrix} I & B^{-1}N & B^{-1}b \\ 0 & c_N^\top - c_B^\top B^{-1}N & f - c_B^\top B^{-1}b \end{pmatrix}$$
$$\equiv \begin{pmatrix} I & B^{-1}N & x_B^* \\ 0 & r_N^\top & f - c_B^\top x_B^* \end{pmatrix}.$$

The above is referred to as a *reduced tableau*. Note that the entries in the last row corresponding to the basic variables x_B are all zero (see the entries under the matrix I).

(2) Check condition (3.8), i.e., $r_N \geqslant 0$. If it holds, then x^* is optimal, and we are done with the optimal objective function value

$$f = c_B^\top x_B^* = c^\top x^*.$$

If (3.8) fails, we choose a $q \in \Lambda_N$ such that (3.11) holds, according to the largest reduction cost rule.

(3) Check condition (3.12). If it holds, then LP (1.1) is unbounded below. If (3.12) fails, then we define the minimum ratio λ by (3.19), and define x^{**} by (3.16), following the minimum ratio with smallest index rule. Such an x^{**} should be another basic feasible solution with a smaller objective function value than that at x^*. Then return to (1).

Practically, we usually omit f in the tableau. Therefore, we will have the following form of the procedure:

$$\begin{pmatrix} B & N & b \\ c_B^\mathsf{T} & c_N^\mathsf{T} & 0 \end{pmatrix} \to \begin{pmatrix} I & B^{-1}N & B^{-1}b \\ 0 & c_N^\mathsf{T} - c_B^\mathsf{T}B^{-1}N & -c_B^\mathsf{T}B^{-1}b \end{pmatrix}.$$

We now repeat Example 3.4 by tableaus.

$$\begin{pmatrix} -1 & \boxed{1} & 1 & 0 & 0 & 5 \\ 1 & 3 & 0 & 1 & 0 & 35 \\ 1 & 0 & 0 & 0 & 1 & 20 \\ -2 & \boxed{-3} & 0 & 0 & 0 & 0 \end{pmatrix} \to \begin{pmatrix} -1 & 1 & 1 & 0 & 0 & 5 \\ \boxed{4} & 0 & -3 & 1 & 0 & 20 \\ 1 & 0 & 0 & 0 & 1 & 20 \\ \boxed{-5} & 0 & 3 & 0 & 0 & 15 \end{pmatrix}$$

$$\to \begin{pmatrix} -1 & 1 & 1 & 0 & 0 & 5 \\ \boxed{1} & 0 & -\frac{3}{4} & \frac{1}{4} & 0 & 5 \\ 1 & 0 & 0 & 0 & 1 & 20 \\ \boxed{-5} & 0 & 3 & 0 & 0 & 15 \end{pmatrix} \to \begin{pmatrix} 0 & 1 & \frac{1}{4} & \frac{1}{4} & 0 & 10 \\ 1 & 0 & -\frac{3}{4} & \frac{1}{4} & 0 & 5 \\ 0 & 0 & \boxed{\frac{3}{4}} & -\frac{1}{4} & 1 & 15 \\ 0 & 0 & \boxed{-\frac{3}{4}} & \frac{5}{4} & 0 & 40 \end{pmatrix}$$

$$\to \begin{pmatrix} 0 & 1 & \frac{1}{4} & \frac{1}{4} & 0 & 10 \\ 1 & 0 & -\frac{3}{4} & \frac{1}{4} & 0 & 5 \\ 0 & 0 & \boxed{1} & -\frac{1}{3} & \frac{4}{3} & 20 \\ 0 & 0 & \boxed{-\frac{3}{4}} & \frac{5}{4} & 0 & 40 \end{pmatrix} \to \begin{pmatrix} 0 & 1 & 0 & \frac{1}{3} & -\frac{1}{3} & 5 \\ 1 & 0 & 0 & 0 & 1 & 20 \\ 0 & 0 & 1 & -\frac{1}{3} & \frac{4}{3} & 20 \\ 0 & 0 & 0 & 1 & 1 & 55 \end{pmatrix}.$$

Thus, the optimal solution is given by

$$x_1 = 20, \quad x_2 = 5, \quad x_3 = 20, \quad x_4 = x_5 = 0,$$

with the optimal objective value

$$c^\mathsf{T}x = -55.$$

Let us make some observations. We look at the first tableau. The basic variables are x_3, x_4, x_5, and the corresponding entries in the last row are already zero. (If they are not zero yet, we need to eliminate them by row operations first.) Thus, the entries of the last row corresponding to the non-basic variables (x_1, x_2) represent the respective cost reductions. Hence, according to the largest reduction cost rule, we select the second column (i.e., select x_2 to enter the basis). Then in that column, we select the first entry which has the minimum ratio (which determines that x_3 will leave the basis). The entry 1 at the position $(1, 2)$ is selected, which is called a *pivot*. The next step is to eliminate all the other non-zero entries in that column. We see from the second tableau that the new basic variables are x_2, x_4, x_5 and the new non-basic variables are x_1, x_3. If at that moment, the last row were non-negative, we are done. Otherwise, we repeat the same procedure to find the next pivot.

Now, let us present another example.

Example 3.7. Consider

$$
\begin{cases}
\min & -3x_1 - x_2, \\
\text{subject to} & \begin{cases} 2x_1 + x_2 \leqslant 6, \\ x_1 + 3x_2 \leqslant 9, \\ x_1, x_2 \geqslant 0. \end{cases}
\end{cases}
$$

We introduce slack variables x_3, x_4 so that the above becomes the following standard LP problem:

$$
\begin{cases}
\min & -3x_1 - x_2, \\
\text{subject to} & \begin{cases} 2x_1 + x_2 + x_3 = 6, \\ x_1 + 3x_2 + x_4 = 9, \\ x_1, x_2, x_3, x_4 \geqslant 0. \end{cases}
\end{cases}
$$

For this problem, $(0, 0, 6, 9)$ is an obvious basic feasible solution with x_3 and x_4 being the basic variables. Now, we use tableau to solve the problem.

$$
\begin{pmatrix} \boxed{2} & 1 & 1 & 0 & 6 \\ 1 & 3 & 0 & 1 & 9 \\ \boxed{-3} & -1 & 0 & 0 & 0 \end{pmatrix} \rightarrow \begin{pmatrix} \boxed{1} & \frac{1}{2} & \frac{1}{2} & 0 & 3 \\ 1 & 3 & 0 & 1 & 9 \\ \boxed{-3} & -1 & 0 & 0 & 0 \end{pmatrix}
$$

$$
\rightarrow \begin{pmatrix} 1 & \frac{1}{2} & \frac{1}{2} & 0 & 3 \\ 0 & \frac{5}{2} & -\frac{1}{2} & 1 & 6 \\ 0 & \frac{1}{2} & \frac{3}{2} & 0 & 9 \end{pmatrix}.
$$

In the last tableau above, the basic variables are x_1 and x_4 and the basic feasible solution is given by $(3, 0, 0, 6)$. Since the reduced cost vector $(\frac{1}{2}, \frac{3}{2})$ is positive, $(x_1, x_2) = (3, 0)$ is an optimal solution.

3.3   *Phase I*

Sometimes, the initial basic feasible solution is not obviously given. In this case, we need to introduce an auxiliary problem which can help us to find a basic feasible solution. Let us make this precise. Again, consider LP (1.1). We introduce an *artificial variable* $z = (z_1, \cdots, z_m)^\top \in \mathbb{R}^m$ and consider the following LP:

$$\begin{cases} \min \mathbf{1}^\top z, \\ \text{subject to } (A, I_m) \begin{pmatrix} x \\ z \end{pmatrix} = b, \quad \begin{pmatrix} x \\ z \end{pmatrix} \geq 0, \end{cases} \tag{3.25}$$

where $\mathbf{1} = (1, 1, \cdots, 1)^\top \in \mathbb{R}^m$. This is called the *Phase I problem*, or *Phase I simplex method*. For this problem, we have the following result.

Proposition 3.8. *Problem* (1.1) *admits a basic feasible solution if and only if Problem* (3.25) *admits an optimal solution with the optimal objective function value 0.*

Proof. Suppose Problem (1.1) has a basic feasible solution x^*. Then $\begin{pmatrix} x^* \\ 0 \end{pmatrix}$ is the optimal feasible solution to Problem (3.25), with the objective function value 0.

Conversely, if $\begin{pmatrix} x^* \\ z^* \end{pmatrix}$ is an optimal feasible solution to Problem (3.25) with optimal objective function value 0, then, it is necessary that $z^* = 0$. Hence, x^* is feasible to Problem (1.1), which means that the feasible set F of LP (1.1) is non-empty. Hence, by Theorem 2.16, we have at least one basic feasible solution to PL (1.1). $\qquad\square$

Practically, we may write the tableau of the Phase I problem as follows:

$$\begin{pmatrix} A & I & b \\ 0 & \mathbf{1}^\top & 0 \end{pmatrix}.$$

Clearly, for this problem, $\begin{pmatrix} 0 \\ b \end{pmatrix}$ is a basic feasible solution with z being the basic variable. Therefore, according to Phase II simplex method, we need first to eliminate the entry $\mathbf{1}^\top$ in the last row by multiplying a suitable matrix:

$$\begin{pmatrix} I & 0 \\ -\mathbf{1}^\top & 1 \end{pmatrix} \begin{pmatrix} A & I & b \\ 0 & \mathbf{1}^\top & 0 \end{pmatrix} = \begin{pmatrix} A & I & b \\ -\mathbf{1}^\top A & 0 & -\mathbf{1}^\top b \end{pmatrix}.$$

Then by Phase II simplex method, we solve the above which will give us a basic feasible solution to the original problem. Let us now look at an example.

Example 3.9. Reconsider Example 1.5 here:

$$\begin{cases} \min \quad x_1 + x_2, \\ \text{subject to} \quad \begin{cases} x_1 + 2x_2 \geqslant 4, \\ -x_1 + x_2 \leqslant 1, \\ 2x_1 + x_2 \leqslant 4, \\ x_1, x_2 \geqslant 0. \end{cases} \end{cases}$$

Let us first transform the problem into a standard form. To this end, we introduce slack variables:

$$\begin{cases} \min \quad x_1 + x_2, \\ \text{subject to} \quad \begin{cases} x_1 + 2x_2 - x_3 = 4, \\ -x_1 + x_2 + x_4 = 1, \\ 2x_1 + x_2 + x_5 = 4, \\ x_1, x_2, x_3, x_4, x_5 \geqslant 0. \end{cases} \end{cases}$$

It is not clear about the first set of the basic variables. Therefore, we introduce artificial variables $z_1, z_2, z_3 \geqslant 0$ and consider the following auxiliary problem:

$$\begin{cases} \min \quad z_1 + z_2 + z_3, \\ \text{subject to} \quad \begin{cases} x_1 + 2x_2 - x_3 + z_1 = 4, \\ -x_1 + x_2 + x_4 + z_2 = 1, \\ 2x_1 + x_2 + x_5 + z_3 = 4, \\ x_1, x_2, x_3, x_4, x_5, z_1, z_2, z_3 \geqslant 0. \end{cases} \end{cases}$$

Thus,

$$\begin{pmatrix} A & I & b \\ 0 & \mathbf{1}^\top & 0 \end{pmatrix} \equiv \begin{pmatrix} 1 & 2 & -1 & 0 & 0 & 1 & 0 & 0 & 4 \\ -1 & 1 & 0 & 1 & 0 & 0 & 1 & 0 & 1 \\ 2 & 1 & 0 & 0 & 1 & 0 & 0 & 1 & 4 \\ 0 & 0 & 0 & 0 & 0 & 1 & 1 & 1 & 0 \end{pmatrix}$$

$$\rightarrow \begin{pmatrix} 1 & 2 & -1 & 0 & 0 & 1 & 0 & 0 & 4 \\ -1 & 1 & 0 & 1 & 0 & 0 & 1 & 0 & 1 \\ 2 & 1 & 0 & 0 & 1 & 0 & 0 & 1 & 4 \\ -2 & -4 & 1 & -1 & -1 & 0 & 0 & 0 & -9 \end{pmatrix}.$$

This step is to eliminate the entries 1 in the last row since z_1, z_2, z_3 are chosen as basic variables. Having finished the current step, the rest of procedure is a Phase II simplex method:

$$
\begin{pmatrix}
1 & 2 & -1 & 0 & 0 & 1 & 0 & 0 & 4 \\
-1 & \boxed{1} & 0 & 1 & 0 & 0 & 1 & 0 & 1 \\
2 & 1 & 0 & 0 & 1 & 0 & 0 & 1 & 4 \\
-2 & \boxed{-4} & 1 & -1 & -1 & 0 & 0 & 0 & -9
\end{pmatrix}
\rightarrow
\begin{pmatrix}
\boxed{3} & 0 & -1 & -2 & 0 & 1 & -2 & 0 & 2 \\
-1 & 1 & 0 & 1 & 0 & 0 & 1 & 0 & 1 \\
3 & 0 & 0 & -1 & 1 & 0 & -1 & 1 & 3 \\
\boxed{-6} & 0 & 1 & 3 & -1 & 0 & 4 & 0 & -5
\end{pmatrix}
$$

$$
\rightarrow
\begin{pmatrix}
\boxed{1} & 0 & -\frac{1}{3} & -\frac{2}{3} & 0 & \frac{1}{3} & -\frac{2}{3} & 0 & \frac{2}{3} \\
-1 & 1 & 0 & 1 & 0 & 0 & 1 & 0 & 1 \\
3 & 0 & 0 & -1 & 1 & 0 & -1 & 1 & 3 \\
\boxed{-6} & 0 & 1 & 3 & -1 & 0 & 4 & 0 & -5
\end{pmatrix}
\rightarrow
\begin{pmatrix}
1 & 0 & -\frac{1}{3} & -\frac{2}{3} & 0 & \frac{1}{3} & -\frac{2}{3} & 0 & \frac{2}{3} \\
0 & 1 & -\frac{1}{3} & \frac{1}{3} & 0 & \frac{1}{3} & \frac{1}{3} & 0 & \frac{5}{3} \\
0 & 0 & \boxed{1} & 1 & 1 & -1 & 1 & 1 & 1 \\
0 & 0 & \boxed{-1} & -1 & -1 & 2 & 0 & 0 & -1
\end{pmatrix}
$$

$$
\rightarrow
\begin{pmatrix}
1 & 0 & 0 & -\frac{1}{3} & \frac{1}{3} & 0 & -\frac{1}{3} & \frac{1}{3} & 1 \\
0 & 1 & 0 & \frac{2}{3} & \frac{1}{3} & 0 & \frac{2}{3} & \frac{1}{3} & 2 \\
0 & 0 & 1 & 1 & 1 & -1 & 1 & 1 & 1 \\
0 & 0 & 0 & 0 & 0 & 1 & 1 & 1 & 0
\end{pmatrix}.
$$

This finishes the Phase I simplex procedure. The selected basic variables are x_1, x_2, x_3, and the resulting tableau for the original problem becomes

$$
\begin{pmatrix}
1 & 0 & 0 & -\frac{1}{3} & \frac{1}{3} & 1 \\
0 & 1 & 0 & \frac{2}{3} & \frac{1}{3} & 2 \\
0 & 0 & 1 & 1 & 1 & 1 \\
1 & 1 & 0 & 0 & 0 & 0
\end{pmatrix}
\rightarrow
\begin{pmatrix}
1 & 0 & 0 & -\frac{1}{3} & \frac{1}{3} & 1 \\
0 & 1 & 0 & \frac{2}{3} & \frac{1}{3} & 2 \\
0 & 0 & 1 & 1 & 1 & 1 \\
0 & 0 & 0 & -\frac{1}{3} & -\frac{2}{3} & -3
\end{pmatrix}.
$$

Then we apply the Phase II procedure again to get the following:

$$
\begin{pmatrix}
1 & 0 & 0 & -\frac{1}{3} & \frac{1}{3} & 1 \\
0 & 1 & 0 & \frac{2}{3} & \frac{1}{3} & 2 \\
0 & 0 & 1 & 1 & \boxed{1} & 1 \\
0 & 0 & 0 & -\frac{1}{3} & \boxed{-\frac{2}{3}} & -3
\end{pmatrix}
\rightarrow
\begin{pmatrix}
1 & 0 & -\frac{1}{3} & -\frac{2}{3} & 0 & \frac{2}{3} \\
0 & 1 & -\frac{1}{3} & \frac{1}{3} & 0 & \frac{5}{3} \\
0 & 0 & 1 & 1 & 1 & 1 \\
0 & 0 & \frac{2}{3} & \frac{1}{3} & 0 & -\frac{7}{3}
\end{pmatrix}.
$$

Thus, the minimum is at $x_1 = \frac{2}{3}$ and $x_2 = \frac{5}{3}$, with the optimal objective function value $\frac{7}{3}$, which coincides with the conclusion of Example 1.5.

3.4   *Endless cycling**

It is possible that the minimum ratio λ defined by (3.19) is zero. This happens when the basic feasible solution x^* is degenerate, and $\xi_j^q < 0$ only for some $j \in \Lambda_B$ such that $x_j^* = 0$. When such a situation happens, $x^{**} = x^*$. But the bases are different. Hence, we may keep selecting pivot and continuing the procedure. However, in the case that there are more than one $q \in \Lambda_N$ such that the most negative cost reduction is attained, and there are more than one $p \in \Lambda_B$ such that the minimum ratio is attained, if we use the largest reduction cost rule for entering the basis and the minimum ratio with smallest index rule for the leaving basis, then the algorithm could end up with an endless cycling. The following example is due to E. M. L. Beale (1955).

Example 3.10. (Beale, 1955) Consider the following LP:

$$\begin{cases} \min & -\dfrac{3}{4}x_4 + 20x_5 - \dfrac{1}{2}x_6 + 6x_7, \\ & \\ \text{subject to} & \begin{cases} x_1 + \dfrac{1}{4}x_4 - 8x_5 - x_6 + 9x_7 = 0, \\ x_2 + \dfrac{1}{2}x_4 - 12x_5 - \dfrac{1}{2}x_6 + 3x_7 = 0, \\ x_3 + x_6 = 1, \\ x_1, x_2, x_3, x_4, x_5, x_6, x_7 \geqslant 0. \end{cases} \end{cases}$$

We use tableaus to carry out the simplex iterations, following the largest reduction cost rule and minimum ratio with smallest index rule. We start with the basis $\{x_1, x_2, x_3\}$ with the following tableau (which is already in its reduced form):

$$(T1) \qquad \begin{pmatrix} 1 & 0 & 0 & \boxed{\tfrac{1}{4}} & -8 & -1 & 9 & 0 \\ 0 & 1 & 0 & \tfrac{1}{2} & -12 & -\tfrac{1}{2} & 3 & 0 \\ 0 & 0 & 1 & 0 & 0 & 1 & 0 & 1 \\ 0 & 0 & 0 & \boxed{-\tfrac{3}{4}} & 20 & -\tfrac{1}{2} & 6 & 0 \end{pmatrix}.$$

By the largest reduction cost rule, we select column 4. Now, there are two indices (1 and 2) for which the minimum ratio 0 is attained. Choosing the one with smaller index. Thus, row 1 is selected. We continue to get next tableaus:

$$(T2) \qquad \begin{pmatrix} 4 & 0 & 0 & \boxed{1} & -32 & -4 & 36 & 0 \\ 0 & 1 & 0 & \tfrac{1}{2} & -12 & -\tfrac{1}{2} & 3 & 0 \\ 0 & 0 & 1 & 0 & 0 & 1 & 0 & 1 \\ 0 & 0 & 0 & -\tfrac{3}{4} & 20 & -\tfrac{1}{2} & 6 & 0 \end{pmatrix} \qquad \text{(normalize row 1)}$$

$$(T3) \qquad \begin{pmatrix} 4 & 0 & 0 & 1 & -32 & -4 & 36 & 0 \\ -2 & 1 & 0 & 0 & \boxed{4} & \tfrac{3}{2} & -15 & 0 \\ 0 & 0 & 1 & 0 & 0 & 1 & 0 & 1 \\ 3 & 0 & 0 & 0 & \boxed{-4} & -\tfrac{7}{2} & 33 & 0 \end{pmatrix} \qquad \begin{array}{l} \text{(elimination in column 4,} \\ \text{choose column 5 and row 2)} \end{array}$$

The current basis is $\{x_2, x_3, x_4\}$. Again, using the largest reduction cost rule, we select column 5. The minimum ratio is 0 which is attained by one index (i.e., 2). Thus, row 2 is selected. We continue the procedure:

$(T4)$
$$\begin{pmatrix} 4 & 0 & 0 & 1 & -32 & -4 & 36 & 0 \\ -\frac{1}{2} & \frac{1}{4} & 0 & 0 & \boxed{1} & \frac{3}{8} & -\frac{15}{4} & 0 \\ 0 & 0 & 1 & 0 & 0 & 1 & 0 & 1 \\ 3 & 0 & 0 & 0 & \boxed{-4} & -\frac{7}{2} & 33 & 0 \end{pmatrix}$$
(normalize row 2)

$(T5)$
$$\begin{pmatrix} -12 & 8 & 0 & 1 & 0 & \boxed{8} & -84 & 0 \\ -\frac{1}{2} & \frac{1}{4} & 0 & 0 & 1 & \frac{3}{8} & -\frac{15}{4} & 0 \\ 0 & 0 & 1 & 0 & 0 & 1 & 0 & 1 \\ 1 & 1 & 0 & 0 & 0 & \boxed{-2} & 18 & 0 \end{pmatrix}$$
(elimination in column 5, choose column 6 and row 1)

The current basis is $\{x_3, x_4, x_5\}$. Again, we select column 6 and since the minimum ratio 0 is attained by two indices (4 and 5), we choose the one with smaller index. Thus, row 1 is selected. We continue:

$(T6)$
$$\begin{pmatrix} -\frac{3}{2} & 1 & 0 & \frac{1}{8} & 0 & \boxed{1} & -\frac{21}{2} & 0 \\ -\frac{1}{2} & \frac{1}{4} & 0 & 0 & 1 & \frac{3}{8} & -\frac{15}{4} & 0 \\ 0 & 0 & 1 & 0 & 0 & 1 & 0 & 1 \\ 1 & 1 & 0 & 0 & 0 & \boxed{-2} & 18 & 0 \end{pmatrix}$$
(normalize row 1)

$(T7)$
$$\begin{pmatrix} -\frac{3}{2} & 1 & 0 & \frac{1}{8} & 0 & 1 & -\frac{21}{2} & 0 \\ \frac{1}{16} & -\frac{1}{8} & 0 & -\frac{3}{64} & 1 & 0 & \boxed{\frac{3}{16}} & 0 \\ \frac{3}{2} & -1 & 1 & -\frac{1}{8} & 0 & 0 & \frac{21}{2} & 1 \\ -2 & 3 & 0 & \frac{1}{4} & 0 & 0 & \boxed{-3} & 0 \end{pmatrix}$$
(elimination in column 6, choose column 7 and row 2)

The current basis is $\{x_3, x_5, x_6\}$. We continue:

$(T8)$
$$\begin{pmatrix} -\frac{3}{2} & 1 & 0 & \frac{1}{8} & 0 & 1 & -\frac{21}{2} & 0 \\ \frac{1}{3} & -\frac{2}{3} & 0 & -\frac{1}{4} & \frac{16}{3} & 0 & \boxed{1} & 0 \\ \frac{3}{2} & -1 & 1 & -\frac{1}{8} & 0 & 0 & \frac{21}{2} & 1 \\ -2 & 3 & 0 & \frac{1}{4} & 0 & 0 & \boxed{-3} & 0 \end{pmatrix}$$
(normalize row 2)

$(T9)$
$$\begin{pmatrix} \boxed{2} & -6 & 0 & -\frac{5}{2} & 56 & 1 & 0 & 0 \\ \frac{1}{3} & -\frac{2}{3} & 0 & -\frac{1}{4} & \frac{16}{3} & 0 & 1 & 0 \\ -2 & 6 & 1 & \frac{5}{2} & -56 & 0 & 0 & 1 \\ \boxed{-1} & 1 & 0 & -\frac{1}{2} & 16 & 0 & 0 & 0 \end{pmatrix}$$
(elimination in column 7, choose column 1 and row 3)

The current basis is $\{x_3, x_6, x_7\}$. We continue:

$$(T10) \qquad \begin{pmatrix} \boxed{1} & -3 & 0 & -\frac{5}{4} & 28 & \frac{1}{2} & 0 & 0 \\ \frac{1}{3} & -\frac{2}{3} & 0 & -\frac{1}{4} & \frac{16}{3} & 0 & 1 & 0 \\ -2 & 6 & 1 & \frac{5}{2} & -56 & 0 & 0 & 1 \\ \boxed{-1} & 1 & 0 & -\frac{1}{2} & 16 & 0 & 0 & 0 \end{pmatrix} \qquad \text{(normalize row 1)}$$

$$(T11) \qquad \begin{pmatrix} 1 & -3 & 0 & -\frac{5}{4} & 28 & \frac{1}{2} & 0 & 0 \\ 0 & \boxed{\frac{1}{3}} & 0 & \frac{1}{6} & -4 & -\frac{1}{6} & 1 & 0 \\ 0 & 0 & 1 & 0 & 0 & 1 & 0 & 1 \\ 0 & \boxed{-2} & 0 & -\frac{7}{4} & 44 & \frac{1}{2} & 0 & 0 \end{pmatrix} \qquad \begin{array}{l}\text{(elimination in column 1,} \\ \text{choose column 2 and row 2)}\end{array}$$

The current basis is $\{x_1, x_3, x_7\}$. We select column 2 and row 2 according our rules. We continue:

$$(T12) \qquad \begin{pmatrix} 1 & -3 & 0 & -\frac{5}{4} & 28 & \frac{1}{2} & 0 & 0 \\ 0 & \boxed{1} & 0 & \frac{1}{2} & -12 & -\frac{1}{2} & 3 & 0 \\ 0 & 0 & 1 & 0 & 0 & 1 & 0 & 1 \\ 0 & \boxed{-2} & 0 & -\frac{7}{4} & 44 & \frac{1}{2} & 0 & 0 \end{pmatrix} \qquad \text{(normalize column 2)}$$

$$(T13) \qquad \begin{pmatrix} 1 & 0 & 0 & \boxed{\frac{1}{4}} & -8 & -1 & 9 & 0 \\ 0 & 1 & 0 & \frac{1}{2} & -12 & -\frac{1}{2} & 3 & 0 \\ 0 & 0 & 1 & 0 & 0 & 1 & 0 & 1 \\ 0 & 0 & 0 & \boxed{-\frac{3}{4}} & 20 & -\frac{1}{2} & 6 & 0 \end{pmatrix} \qquad \text{(elimination in column 2)}$$

We note that $(T13)$ coincides with $(T1)$. Hence, the simplex iteration gets into a cycling.

To prevent an endless cycling, people introduced several interesting rules for picking up the decision variables to leave and to enter the basis, respectively. A simple one is called *Bland's Rule* which can be stated as follows:

(i) Among all non-basic variables in negative reduced costs, choose the one with the smallest index to enter the basis;

(ii) When there is a tie in the minimum ratios, choose the basic variable with the smallest index to leave the basis.

One can prove that with Bland's rule, there will be no endless cycling. We now apply Bland's rule to the above example.

We start from the basis $\{x_1, x_2, x_3\}$, and keep $(T1)$–$(T7)$. For convenience, we copy it as $(T7')$ here:

$$(T7')\quad
\begin{pmatrix}
-\frac{3}{2} & 1 & 0 & \frac{1}{8} & 0 & 1 & -\frac{21}{2} & 0 \\
\boxed{\frac{1}{16}} & -\frac{1}{8} & 0 & -\frac{3}{64} & 1 & 0 & \frac{3}{16} & 0 \\
\frac{3}{2} & -1 & 1 & -\frac{1}{8} & 0 & 0 & \frac{21}{2} & 1 \\
\boxed{-2} & 3 & 0 & \frac{1}{4} & 0 & 0 & -3 & 0
\end{pmatrix}
\quad\begin{array}{l}\text{(elimination in column 6,}\\[2pt]\text{choose column 1 and row 2)}\end{array}$$

Note that we select column 1 in $(T7')$ instead of column 7 in $(T7)$, because we choose the column with a negative reduced cost (not necessarily the most negative) and with the smallest index. Now, we continue.

$$(T8')\quad
\begin{pmatrix}
-\frac{3}{2} & 1 & 0 & \frac{1}{8} & 0 & 1 & -\frac{21}{2} & 0 \\
\boxed{1} & -2 & 0 & -\frac{3}{4} & 16 & 0 & 3 & 0 \\
\frac{3}{2} & -1 & 1 & -\frac{1}{8} & 0 & 0 & \frac{21}{2} & 1 \\
\boxed{-2} & 3 & 0 & \frac{1}{4} & 0 & 0 & -3 & 0
\end{pmatrix}
\quad\text{(normalize row 2)}$$

$$(T9')\quad
\begin{pmatrix}
0 & -2 & 0 & -1 & 24 & 1 & -6 & 0 \\
1 & -2 & 0 & -\frac{3}{4} & 16 & 0 & 3 & 0 \\
0 & \boxed{2} & 1 & 1 & -24 & 0 & 6 & 1 \\
0 & \boxed{-1} & 0 & -\frac{5}{4} & 32 & 0 & 3 & 0
\end{pmatrix}
\quad\begin{array}{l}\text{(elimination in column 1,}\\[2pt]\text{choose column 2 and row 3)}\end{array}$$

$$(T10')\quad
\begin{pmatrix}
0 & -2 & 0 & -1 & 24 & 1 & -6 & 0 \\
1 & -2 & 0 & -\frac{3}{4} & 16 & 0 & 3 & 0 \\
0 & \boxed{1} & \frac{1}{2} & \frac{1}{2} & -12 & 0 & 3 & \frac{1}{2} \\
0 & \boxed{-1} & 0 & -\frac{5}{4} & 32 & 0 & 3 & 0
\end{pmatrix}
\quad\text{(normalize row 3)}$$

$$(T11')\quad
\begin{pmatrix}
0 & 0 & 1 & 0 & 0 & 1 & 0 & 1 \\
1 & 0 & 1 & \frac{1}{4} & -8 & 0 & 9 & 1 \\
0 & 1 & \frac{1}{2} & \boxed{\frac{1}{2}} & -12 & 0 & 3 & \frac{1}{2} \\
0 & 0 & \frac{1}{2} & \boxed{-\frac{3}{4}} & 20 & 0 & 6 & \frac{1}{2}
\end{pmatrix}
\quad\begin{array}{l}\text{(elimination in column 2,}\\[2pt]\text{choose column 4 and row 2)}\end{array}$$

Note that at this moment, we have got out of the "cycle". The minimum ratios in the above are not zero anymore. We continue the above tableaus:

$$(T12')\quad
\begin{pmatrix}
0 & 0 & 1 & 0 & 0 & 1 & 0 & 1 \\
1 & 0 & 1 & \frac{1}{4} & -8 & 0 & 9 & 1 \\
0 & 2 & 1 & \boxed{1} & -24 & 0 & 6 & 1 \\
0 & 0 & \frac{1}{2} & \boxed{-\frac{3}{4}} & 20 & 0 & 6 & \frac{1}{2}
\end{pmatrix}
\quad\text{(normalize row 3).}$$

$$(T13') \quad \begin{pmatrix} 0 & 0 & 1 & 0 & 0 & 1 & 0 & 1 \\ 1 & -\frac{1}{2} & \frac{3}{4} & 0 & -2 & 0 & \frac{15}{2} & \frac{3}{4} \\ 0 & 2 & 1 & 1 & -24 & 0 & 6 & 1 \\ 0 & \frac{3}{2} & \frac{5}{4} & 0 & 2 & 0 & \frac{21}{2} & \frac{5}{4} \end{pmatrix} \qquad \text{(elimination in column 4)}$$

We see that the last row is non-negative. Thus, the optimal solution is given by

$$x_1 = \frac{3}{4}, \ x_4 = 1, \ x_6 = 1,$$

with the optimal objective function value

$$-\frac{3}{4}x_4 + 20x_5 - \frac{1}{2}x_6 + 6x_7 = -\frac{5}{4}.$$

Exercises

1. Find all basic feasible solutions and the corresponding reduced cost vectors for each of the following triples (A, b, c), from which determine a solution to the corresponding LP problem:

(i) $A = \begin{pmatrix} 1 & 2 & 3 \\ -1 & 1 & 0 \end{pmatrix}$, $b = \begin{pmatrix} 4 \\ 2 \end{pmatrix}$, $c^\top = (-1, 2, 1)$;

(ii) $A = \begin{pmatrix} 1 & 2 & 0 & 3 \\ 0 & 1 & 0 & 4 \end{pmatrix}$, $b = \begin{pmatrix} 1 \\ 1 \end{pmatrix}$, $c^\top = (1, 2, 3, 4)$.

2. Solve the following LP problem by simplex method:

$$\begin{cases} \min & x_1 - x_2 + x_3, \\ \text{subject to} & \begin{cases} x_1 - x_2 = 2, \\ x_2 - x_3 = 4, \\ x_1, x_2, x_3 \geqslant 0. \end{cases} \end{cases}$$

3. Solving the following LP problem by simplex method:

$$\begin{cases} \min & -x_1 - x_2, \\ \text{subject to} & \begin{cases} x_1 + 2x_2 - x_3 = 4, \\ -x_1 + x_2 + x_4 = 1, \\ 2x_1 + x_2 + x_5 = 4, \\ x_1, x_2, x_3, x_4, x_5 \geqslant 0. \end{cases} \end{cases}$$

4. Solve the following LP problem by simplex method:

$$\begin{cases} \min x_1 + 2x_2, \\ \text{subject to} \quad \begin{cases} x_1 + x_2 \geqslant 1, \\ x_1 + x_2 \leqslant 2, \\ x_1, x_2 \geqslant 0. \end{cases} \end{cases}$$

5. Solve the following LP problem by simplex method:

$$\begin{cases} \min -x_1 - 2x_2, \\ \text{subject to} \quad \begin{cases} x_1 + x_2 \geqslant 1, \\ x_1 + x_2 \leqslant 2, \\ x_1, x_2 \geqslant 0. \end{cases} \end{cases}$$

4   Sensitivity Analysis

In practice, it often happens that the data/measurements have errors, which leads to the inaccuracy of the model. Then a natural question is how such kinds of errors affect the optimality of the optimal solution for the inaccurate model? In other words, suppose we find an optimal solution x^* to an LP problem, is it still an optimal solution to the LP problem if the data have been perturbed (within a range)? We may further ask what happens if a decision variable is added or removed? A constraint is added or removed? We now look at these questions in some details. Let us consider the standard form LP problem (1.1), and let $x^* \in \mathcal{E}(F)$ be an optimal solution of (1.1) with the basic decomposition (3.1)–(3.2). Thus, x^* is given by (3.5).

Case 1. The right hand side is perturbed.

Suppose $\bar{b} \in \mathbb{R}^m$ is given and we consider the following perturbed LP problem:

$$\begin{cases} \min \quad c^\top x, \\ \text{subject to} \quad Ax = b + \mu\bar{b}, \qquad x \geqslant 0, \end{cases} \tag{4.1}$$

where $\mu \in \mathbb{R}$. Now, we define

$$x(\mu) = \begin{pmatrix} B^{-1}(b + \mu\bar{b}) \\ 0 \end{pmatrix}.$$

Then $x(\mu)$ satisfies the equality constraint in (4.1). Note that by Proposition 3.2, when x^* is non-degenerate, the optimality of x^* implies the non-negativity of the reduced cost vector $r_N \equiv c_N - (B^{-1}N)^\top c_B$ which is independent of the right hand side vector of LP problem. Hence, it remains non-negative when b changes to $b + \mu\bar{b}$. Consequently, as long as $x(\mu)$ is feasible, it will be an optimal solution for the perturbed problem. Now, $x(\mu)$ is feasible if and only if

$$x_B(\mu) \equiv B^{-1}b + \mu B^{-1}\bar{b} \geqslant 0.$$

Thus, we need

$$\begin{cases} \mu \leqslant \dfrac{(B^{-1}b)_j}{-(B^{-1}\bar{b})_j}\,, & (B^{-1}\bar{b})_j < 0, \ j \in \Lambda_B, \\[3mm] \mu \geqslant \dfrac{(B^{-1}b)_j}{-(B^{-1}\bar{b})_j}\,, & (B^{-1}\bar{b})_j > 0, \ j \in \Lambda_B. \end{cases}$$

Hence, if we define

$$\begin{cases} \underline{\mu} = \max\left\{ \dfrac{(B^{-1}b)_j}{-(B^{-1}\bar{b})_j} \,\middle|\, (B^{-1}\bar{b})_j > 0, \ j \in \Lambda_B \right\}, \\[3mm] \bar{\mu} = \min\left\{ \dfrac{(B^{-1}b)_j}{-(B^{-1}\bar{b})_j} \,\middle|\, (B^{-1}\bar{b})_j < 0, \ j \in \Lambda_B \right\}, \end{cases} \tag{4.2}$$

then for any $\mu \in [\underline{\mu}, \bar{\mu}]$, $x(\mu)$ will be optimal and the corresponding objective function value is given by

$$c^\top x(\mu) = c^\top x^* + \mu c_B^\top B^{-1}\bar{b}. \tag{4.3}$$

Let us denote $w^* \equiv (B^{-1})^\top c_B$, $\Delta b = \mu\bar{b}$ and $\Delta x = x(\mu) - x^*$. Then Δb is the (perturbation) increment of the right hand side of the equality constraint and Δx is the corresponding increment of the decision variable. With these notations, (4.3) is equivalent to the following:

$$c^\top \Delta x = (w^*)^\top \Delta b. \tag{4.4}$$

If we regard our LP problem as a process of providing different services $(x \geqslant 0)$ to meet a set of customer demand $(Ax = b)$ in the least expensive way (measured by the cost $c^\top x$), then the above means that w_i^* is the *marginal cost* of the providing one additional unit of the i-th demand at the optimal solution x^*. Thus, we call w^* the *marginal prices*, the *shadow prices*, or the *equilibrium prices*.

Let us look at the following example.

Example 4.1. Consider

$$\begin{cases} \min & -3x_1 - x_2, \\ \text{subject to} & \begin{cases} 2x_1 + x_2 + x_3 = 6, \\ x_1 + 3x_2 + x_4 = 9, \\ x_1, x_2, x_3, x_4 \geqslant 0. \end{cases} \end{cases}$$

This problem is solved in Example 3.7. The optimal solution is given by $(3, 0, 0, 6)$, with the basic variable vector $x_B = (x_1, x_4)^\top$ and non-basic variable vector $x_N = (x_2, x_3)^\top$. Hence,

$$B = \begin{pmatrix} 2 & 0 \\ 1 & 1 \end{pmatrix}, \quad N = \begin{pmatrix} 1 & 1 \\ 3 & 0 \end{pmatrix}.$$

$$x_B^* = B^{-1}b = \begin{pmatrix} \frac{1}{2} & 0 \\ -\frac{1}{2} & 1 \end{pmatrix} \begin{pmatrix} 6 \\ 9 \end{pmatrix} = \begin{pmatrix} 3 \\ 6 \end{pmatrix},$$

and

$$r_N^\top = c_N^\top - c_B^\top B^{-1}N = (-1, 0) - (-3, 0) \begin{pmatrix} \frac{1}{2} & 0 \\ -\frac{1}{2} & 1 \end{pmatrix} \begin{pmatrix} 1 & 1 \\ 3 & 0 \end{pmatrix} = \begin{pmatrix} 1 \\ 2 \end{pmatrix}, \frac{3}{2}) > 0.$$

Now, suppose the equality constraint is perturbed as follows:

$$\begin{cases} 2x_1 + x_2 + x_3 = 6 + \mu, \\ x_1 + 3x_2 + x_4 = 9 - 2\mu, \\ x_1, x_2, x_3, x_4 \geqslant 0. \end{cases}$$

Thus, $\bar{b} = (1, -2)^\top$, and we need

$$0 \leqslant x_B(\mu) \equiv B^{-1}b + \mu B^{-1}\bar{b} = \begin{pmatrix} \frac{1}{2} & 0 \\ -\frac{1}{2} & 1 \end{pmatrix} \begin{pmatrix} 6 \\ 9 \end{pmatrix} + \mu \begin{pmatrix} \frac{1}{2} & 0 \\ -\frac{1}{2} & 1 \end{pmatrix} \begin{pmatrix} 1 \\ -2 \end{pmatrix}$$

$$= \begin{pmatrix} 3 \\ 6 \end{pmatrix} + \mu \begin{pmatrix} \frac{1}{2} \\ -\frac{5}{2} \end{pmatrix} = \begin{pmatrix} 3 + \frac{\mu}{2} \\ 6 - \frac{5}{2}\mu \end{pmatrix}.$$

Hence, we require

$$-6 \leqslant \mu \leqslant \frac{12}{5}.$$

Once μ satisfies the above, $x(\mu) = (3 + \frac{\mu}{2}, 0, 0, 6 - \frac{5}{2}\mu)$ will be optimal to the perturbed LP problem. Further, in this case, the shadow price vector is given by

$$w^* = (B^{-1})^\top c_B = \begin{pmatrix} \frac{1}{2} & -\frac{1}{2} \\ 0 & 1 \end{pmatrix} \begin{pmatrix} -3 \\ 0 \end{pmatrix} = \begin{pmatrix} -\frac{3}{2} \\ 0 \end{pmatrix}.$$

Case 2. The cost vector is perturbed.

Suppose the cost vector c is perturbed (in practice, it could be understood as price changes, etc.) to $c + \mu\bar{c}$, where $\bar{c} \in \mathbb{R}^n$ is given, and $\mu \in \mathbb{R}$. We may assume that

$$c + \mu\bar{c} \equiv \begin{pmatrix} c_B \\ c_N \end{pmatrix} + \mu \begin{pmatrix} \bar{c}_B \\ \bar{c}_N \end{pmatrix}.$$

Now, the question is whether x^* remains optimal for small $|\mu|$. To obtain the optimality of x^* with the new cost vector, we need only to look at the corresponding reduced cost vector:

$$r_N(\mu) = (c_N + \mu \bar{c}_N) - (B^{-1}N)^\top (c_B + \mu \bar{c}_B)$$
$$= c_N + (B^{-1}N)^\top c_B + \mu[\bar{c}_N - (B^{-1}N)^\top \bar{c}_B] \equiv r_N + \mu \bar{r}_N.$$

Hence, to guarantee the optimality of x^*, we need only to have

$$r_N + \mu \bar{r}_N \geqslant 0.$$

Now, we set (similar to (4.2))

$$\begin{cases} \underline{\mu} = \max \left\{ \dfrac{r_j}{-\bar{r}_j} \ \Big| \ \bar{r}_j > 0, \ j \in \Lambda_N \right\}, \\ \overline{\mu} = \min \left\{ \dfrac{r_j}{-\bar{r}_j} \ \Big| \ \bar{r}_j < 0, \ j \in \Lambda_N \right\}. \end{cases} \tag{4.5}$$

Then for any $\mu \in [\underline{\mu}, \overline{\mu}]$, x^* remains optimal, with the optimal objective function value

$$(c + \mu \bar{c})^\top x^* = c^\top x^* + \mu \bar{c}^\top x^*. \tag{4.6}$$

Example 4.2. Let us look at the LP problem in Example 4.1 again. Suppose the cost vector is perturbed to the following:

$$c + \mu \bar{c} = (-3 + \mu, -1 - \mu, 0, 0).$$

Then

$$r_N(\mu)^\top = (-1 - \mu, 0) - (-3 + \mu, 0) \begin{pmatrix} \frac{1}{2} & 0 \\ -\frac{1}{2} & 1 \end{pmatrix} \begin{pmatrix} 1 & 1 \\ 3 & 0 \end{pmatrix}$$
$$= (-1 - \mu, 0) + \left(\frac{3 - \mu}{2}, \frac{3 - \mu}{2} \right) = \left(\frac{1 - 3\mu}{2}, \frac{3 - \mu}{2} \right).$$

Hence, as long as

$$\mu \leqslant \frac{1}{3},$$

the original optimal solution $x^* = (3, 0, 0, 6)^\top$ remains optimal.

Case 3. Adding a new decision variable.

This case happens when an additional decision variable becomes available and has to be considered after solving the original LP problem. Let the new decision variable be x_{n+1}. Then the new LP problem takes the following form:

$$\begin{cases} \min & c^\top x + c_{n+1} x_{n+1}, \\ \text{subject to} & Ax + a_{n+1} x_{n+1} = b, \quad x \geqslant 0, \quad x_{n+1} \geqslant 0, \end{cases}$$

where $a_{n+1} \in \mathbb{R}^m$ and $c_{n+1} \in \mathbb{R}$ are given. Clearly, $\bar{x}^* \equiv \begin{pmatrix} x^* \\ 0 \end{pmatrix}$ is a basic feasible solution to the new LP problem. Also, we know that $r_N \geqslant 0$ (due to the non-degeneracy of x^*). Thus, in order $\bar{x}^*$ to be optimal, we need only to check the following condition:

$$r_{n+1} = c_{n+1} - c_B^\top B^{-1} a_{n+1} \geqslant 0. \tag{4.7}$$

If the above holds, then $\bar{x}^*$ is optimal and the new decision variable does not do anything. On the other hand, if (4.7) fails, then we have to use the simplex method starting from $\bar{x}^*$.

Example 4.3. Let us still consider LP problem in Example 4.1. We now add a decision variable x_5 so that the problem becomes

$$\begin{cases} \min & -3x_1 - x_2 - x_5, \\ \text{subject to} & \begin{cases} 2x_1 + x_2 + x_3 + x_5 = 6, \\ x_1 + 3x_2 + x_4 - x_5 = 9, \\ x_1, x_2, x_3, x_4, x_5 \geqslant 0. \end{cases} \end{cases}$$

Then $a_5 = (1, -1)^\top$ and $c_5 = -1$. Thus,

$$r_5 = c_5 - c_B^\top B^{-1} a_5 = -1 - (-3, 0) \begin{pmatrix} \frac{1}{2} & 0 \\ -\frac{1}{2} & 1 \end{pmatrix} \begin{pmatrix} 1 \\ -1 \end{pmatrix} = \frac{1}{2} > 0.$$

Therefore, the solution $(3, 0, 0, 6, 0)$ is optimal with optimal objective function value -9.

On the other hand, if we have the following situation:

$$\begin{cases} \min & -3x_1 - x_2 - 2x_5, \\ \text{subject to} & \begin{cases} 2x_1 + x_2 + x_3 + x_5 = 6, \\ x_1 + 3x_2 + x_4 - x_5 = 9, \\ x_1, x_2, x_3, x_4, x_5 \geqslant 0. \end{cases} \end{cases}$$

Then $a_5 = (1, -1)^\top$ and $c_5 = -2$. Thus,

$$r_5 = c_5 - c_B^\top B^{-1} a_5 = -2 - (-3, 0) \begin{pmatrix} \frac{1}{2} & 0 \\ -\frac{1}{2} & 1 \end{pmatrix} \begin{pmatrix} 1 \\ -1 \end{pmatrix} = -\frac{1}{2} < 0.$$

This implies that the basic feasible solution $(3, 0, 0, 6, 0)$ is not optimal, and we need to use Phase II simplex method to find an optimal solution. In this case, we can obtain the optimal solution $(0, 0, 0, 15, 6)$ with the optimal objective function value -12.

Case 4. Removing a decision variable.

This case happens when a decision variable becomes unavailable and has to be removed from the decision variables. Without loss of generality, we assume that x_n is going to be removed. We start with the (non-degenerate) optimal solution x^*. If $x_n^* = 0$, then $\bar{x}^* \triangleq (x_1^*, \cdots, x_{n-1}^*)^\top$ will be the optimal solution to the new problem. Otherwise, we will have a new LP problem and we could apply the technique developed in the previous sections to solve.

Example 4.4. Again for the problem in Example 4.1, if we remove decision variable x_2 or x_3, the original optimal solution will clearly remain, i.e., the optimal solution for the new problem will be $(3, 0, 6)$. However, if we remove decision variable x_4, then the problem becomes

$$
\begin{cases}
\min & -3x_1 - x_2, \\
\text{subject to} & \begin{cases} 2x_1 + x_2 + x_3 = 6, \\ x_1 + 3x_2 = 9, \\ x_1, x_2, x_3 \geqslant 0. \end{cases}
\end{cases}
$$

In this case, we may directly find

$$
\begin{cases}
x_1 = 9 - 3x_2, \\
x_3 = -12 + 5x_2.
\end{cases}
$$

Thus, the feasible set is given by

$$
F = \left\{ \begin{pmatrix} 9 - 3x_2 \\ x_2 \\ -12 + 5x_2 \end{pmatrix} \;\middle|\; \frac{12}{5} \leqslant x_2 \leqslant 3 \right\},
$$

which compact. Then we are able to obtain two critical points:

$$
x^* = \begin{pmatrix} 0 \\ 3 \\ 3 \end{pmatrix}, \quad x^{**} = \begin{pmatrix} \frac{9}{5} \\ \frac{12}{5} \\ 0 \end{pmatrix},
$$

with

$$
c^\top x^* = -3, \quad c^\top x^{**} = -\frac{39}{5}.
$$

Hence, x^{**} is the optimal solution.

It is possible that when one or more decision variables are removed, the problem might become infeasible. Actually, if we have the following equality constraints:

$$\begin{cases} x_1 + x_2 - x_3 = 1, \\ x_1 + x_2 + 3x_3 = 2, \end{cases}$$

and if x_3 is removed, then the feasible set is empty, and the corresponding LP problem becomes infeasible.

Case 5. Removing/adding a constraint.

Usually, removing a constraint will make the feasible set larger and the optimal objective function value smaller; adding a constraint will make the feasible set smaller and the optimal object function value larger. More than often, one might have to solve a new LP problem.

Example 4.5. We still consider the LP problem in Example 4.1. We first remove the first constraint. Then the problem becomes

$$\begin{cases} \min \quad -3x_1 - x_2, \\ \text{subject to} \begin{cases} x_1 + 3x_2 + x_4 = 9, \\ x_1, x_2, x_4 \geqslant 0. \end{cases} \end{cases}$$

Since $m = 1$, the extreme points are

$$(9, 0, 0), \quad (0, 3, 0), \quad (0, 0, 9).$$

Clearly, the optimal solution is $(9, 0, 0)$ with the optimal objective function value -27 which is smaller than the original one.

Now, we remove the second constraint. Then the problem becomes

$$\begin{cases} \min \quad -3x_1 - x_2, \\ \text{subject to} \begin{cases} 2x_1 + x_2 + x_3 = 6, \\ x_1, x_2, x_3, x_4 \geqslant 0. \end{cases} \end{cases}$$

In this case, the extreme points are

$$(3, 0, 0), \quad (0, 6, 0), \quad (0, 0, 6).$$

Clearly, the optimal solution is $(3, 0, 0)$ with optimal objective function value -9.

From the above, we see that removing different constraints will lead to different optimal solutions, which is pretty natural.

Finally, let us consider the situation that a constraint is added. For example, the original problem might become the following:

$$\begin{cases} \min & -3x_1 - x_2, \\ \text{subject to} & \begin{cases} 2x_1 + x_2 + x_3 = 6, \\ x_1 + 3x_2 + x_4 = 9, \\ x_1 + x_2 = \dfrac{21}{5}, \\ x_1, x_2, x_3, x_4 \geqslant 0. \end{cases} \end{cases}$$

The above LP problem has a unique feasible solution: $(\frac{9}{5}, \frac{12}{5}, 0, 0)$. Therefore, it is the optimal solution with the optimal objective function value $-\frac{39}{5}$ which is larger than the original one which is -9. Furthermore, if we add a constraint so that the problem becomes

$$\begin{cases} \min & -3x_1 - x_2, \\ \text{subject to} & \begin{cases} 2x_1 + x_2 + x_3 = 6, \\ x_1 + 3x_2 + x_4 = 9, \\ -x_1 + 2x_2 - x_3 + x_4 = 4, \\ x_1, x_2, x_3, x_4 \geqslant 0, \end{cases} \end{cases}$$

then the feasible set is empty since the sum of the first and the third equality constraints gives

$$x_1 + 3x_2 + x_4 = 10,$$

which contradicting the second equality constraint. Thus, the problem becomes infeasible. Finally, if a constraint is added so that the problem becomes

$$\begin{cases} \min & -3x_1 - x_2, \\ \text{subject to} & \begin{cases} 2x_1 + x_2 + x_3 = 6, \\ x_1 + 3x_2 + x_4 = 9, \\ -x_1 + 2x_2 - x_3 + x_4 = 3, \\ x_1, x_2, x_3, x_4 \geqslant 0, \end{cases} \end{cases}$$

then the problem is equivalent to the original one, since the sum of the first and the third constraints equals the second one. Therefore, the optimal solution is unchanged.

Exercises

For the following LP problem:

$$\begin{cases} \min & -x_1 - x_2, \\ \text{subject to} & \begin{cases} x_1 + 2x_2 - x_3 = 4, \\ -x_1 + x_2 + x_4 = 1, \\ 2x_1 + x_2 + x_5 = 4, \\ x_1, x_2, x_3, x_4, x_5 \geqslant 0, \end{cases} \end{cases}$$

carry out the following:

(i) Suppose the constraints are perturbed as follows:

$$\begin{cases} x_1 + 2x_2 - x_3 = 4 - \mu, \\ -x_1 + x_2 + x_4 = 1 + 2\mu, \\ 2x_1 + x_2 + x_5 = 4 + \mu, \\ x_1, x_2, x_3, x_4, x_5 \geqslant 0. \end{cases}$$

Find the range of μ so that the corresponding $x(\mu)$ is optimal. Find the shadow price.

(ii) Suppose the cost vector is replaced by the following:

$$c + \mu\bar{c} = (-1 + \mu, -1 + 2\mu, \mu, 2\mu, -\mu).$$

Find the range of μ so that the original optimal solution x^* remains optimal. Calculate the corresponding new optimal objective function value.

(iii) Add a new decision variable x_6 so that the problem becomes

$$\begin{cases} \min & -x_1 - x_2 - x_6, \\ \text{subject to} & \begin{cases} x_1 + 2x_2 - x_3 + x_6 = 4, \\ -x_1 + x_2 + x_4 - 2x_6 = 1, \\ 2x_1 + x_2 + x_5 = 4, \\ x_1, x_2, x_3, x_4, x_5, x_6 \geqslant 0. \end{cases} \end{cases}$$

Find the new optimal solution, compare with the original optimal solution.

(iv) Remove decision variable x_4. Solve the corresponding LP problem.

(v) Remove the first equality constraint and solve the corresponding LP problem.

(vi) Add an equality constraint

$$x_1 - x_2 + x_3 - x_4 + x_5 = 0.$$

Solve the corresponding problem.

5 Duality Theory

In Chapter 4, we have presented the Lagrange duality for convex optimization problems. We now would like to look at the duality theory for the standard form LP (1.1).

Mimicking those in Section 3 of Chapter 4, for LP (1.1), we define

$$\theta(\lambda, \mu) = \inf_{x \in \mathbb{R}^n} \left[c^\top x + \lambda^\top (Ax - b) - \mu^\top x \right]$$
$$= \inf_{x \in \mathbb{R}^n} \left[(c + A^\top \lambda - \mu)^\top x \right] - b^\top \lambda = -b^\top \lambda,$$

provided

$$A^\top(-\lambda) = c - \mu \leqslant c.$$

Thus, according to Section 3 of Chapter 4, we should let

$$\Phi = \left\{ (\lambda, \mu) \in \mathbb{R}^m \times \mathbb{R}^n_+ \mid A^\top(-\lambda) = c - \mu \right\},$$

and the dual problem reads

$$\begin{cases} \max & b^\top(-\lambda), \\ \text{subject to} & (\lambda, \mu) \in \Phi. \end{cases}$$

By letting $w = -\lambda \in \mathbb{R}^m$ and noting $\mu \geqslant 0$, we see that the above problem is equivalent to the following:

$$\begin{cases} \max & b^\top w, \\ \text{subject to} & A^\top w \leqslant c, \quad w \in \mathbb{R}^m. \end{cases} \tag{5.1}$$

Note that in the above, w is unrestricted. Problem (5.1) is called the *dual problem* of the *primal problem* (1.1). We call (1.1) and (5.1) a *primal-dual pair*. Clearly, (5.1) is equivalent to the following:

$$\begin{cases} \min & (-b)^\top w, \\ \text{subject to} & A^\top w \leqslant c, \quad w \in \mathbb{R}^m. \end{cases} \tag{5.2}$$

We introduce $u, v \geqslant 0$ to represent $w = u - v$. Denote $y = (u^\top, v^\top, s^\top)^\top$ with $s \geqslant 0$ being a slack variable. Then the above problem (5.2) is equivalent to the following standard form LP:

$$\begin{cases} \min & (-b^\top, b^\top, 0)y, \\ \text{subject to} & (A^\top, -A^\top, I)y = c, \quad y \geqslant 0. \end{cases} \tag{5.3}$$

For this standard form LP problem, we can further write its dual problem as follows:

$$\begin{cases} \max \quad (-c)^\top w, \\ \text{subject to} \quad \begin{pmatrix} A \\ -A \\ I \end{pmatrix} w \leqslant \begin{pmatrix} -b \\ b \\ 0 \end{pmatrix}, \quad w \in \mathbb{R}^m. \end{cases}$$

By letting $x = -w$, we see that the above is equivalent to LP (1.1). This means that the dual of (5.1) (which is the dual of (1.1)) is (1.1) itself, i.e., the dual of the dual is back to the original problem.

Now, we recall the primal problem of canonical form (1.5) here

$$\begin{cases} \min \quad c^\top x, \\ \text{subject to} \quad Ax \geqslant b, \quad x \geqslant 0. \end{cases} \tag{5.4}$$

It is equivalent to the following standard primal problem:

$$\begin{cases} \min \quad \begin{pmatrix} c \\ 0 \end{pmatrix}^\top \begin{pmatrix} x \\ y \end{pmatrix}, \\ \text{subject to} \quad (A, -I) \begin{pmatrix} x \\ y \end{pmatrix} = b, \\ \qquad\qquad \begin{pmatrix} x \\ y \end{pmatrix} \geqslant 0. \end{cases}$$

By definition, the dual problem of the above reads

$$\begin{cases} \max \quad b^\top w, \\ \begin{pmatrix} A^\top \\ -I \end{pmatrix} w \leqslant \begin{pmatrix} c \\ 0 \end{pmatrix}, \quad w \in \mathbb{R}^m. \end{cases}$$

This is equivalent to the following:

$$\begin{cases} \max \quad b^\top w, \\ \text{subject to} \quad A^\top w \leqslant c, \quad w \geqslant 0. \end{cases} \tag{5.5}$$

We called (5.5) the *dual problem* of canonical form LP problem (1.5). Clearly, (5.5) is equivalent to

$$\begin{cases} \min \quad (-b)^\top w, \\ \text{subject to} \quad (-A)^\top w \geqslant (-c), \quad w \geqslant 0. \end{cases} \tag{5.6}$$

This is again a canonical form LP whose dual looks like the following:

$$\begin{cases} \max \quad (-c)^\top x, \\ \text{subject to} \quad (-A)^\top x \leqslant -b, \quad x \geqslant 0, \end{cases}$$

which is equivalent to (5.4). Formally, if we use the following matrix

$$\begin{pmatrix} A & b \\ c^\top & 0 \end{pmatrix}$$

to represent canonical form LP (5.4), then LP (5.6) is represented by

$$\begin{pmatrix} (-A)^\top & (-c) \\ (-b)^\top & 0 \end{pmatrix} = -\begin{pmatrix} A & b \\ c^\top & 0 \end{pmatrix}^\top.$$

From this, one sees immediately that the dual problem of (5.6) (or (5.5)) is (1.5). This again means that the dual problem of the dual problem is the original primal problem for canonical form LP problems.

In what follows, we consider the primal-dual pair (1.1) and (5.1). First, we have the following result which is a special case of Theorem 3.1 in Chapter 4.

Theorem 5.1. (Weak Duality Principle) *Consider the standard form LP* (1.1) *and its dual LP* (5.1). *Let F_P and F_D be the feasible sets of them, respectively, i.e.,*

$$F_P = \left\{ x \in \mathbb{R}^n \mid Ax = b, \quad x \geqslant 0 \right\},$$
$$F_D = \left\{ w \in \mathbb{R}^m \mid A^\top w \leqslant c \right\}.$$

(i) *If $F_P \neq \varnothing$ and $F_D \neq \varnothing$, then*

$$c^\top x \geqslant b^\top w, \qquad \forall x \in F_P, \ w \in F_D. \tag{5.7}$$

Consequently,

$$\sup_{w \in F_D} b^\top w \leqslant \inf_{x \in F_P} c^\top x. \tag{5.8}$$

(ii) *If $\Phi_D \neq \varnothing$ and*

$$\sup_{w \in F_D} b^\top w = \infty,$$

then $\Phi_P = \varnothing$.

(iii) *If $F_P \neq \varnothing$ and*

$$\inf_{x \in F_P} c^\top x = -\infty,$$

then $F_D = \varnothing$.

(iv) *If there are $\bar{x} \in F_P$ and $\bar{w} \in F_D$ satisfying*

$$c^\top \bar{x} = b^\top \bar{w}, \tag{5.9}$$

then $\bar{x}$ and $\bar{w}$ are optimal solutions of primal LP (1.1) *and dual LP* (5.1), *respectively. In this case, $\bar{w}$ is the shadow price of LP problem* (1.1).

The proof can be copied line-by-line from that of Theorem 3.1 of Chapter 4, and we leave it to the readers.

The above weak duality principal also holds for canonical form LP problem (1.5) and its dual (5.5). We leave the proof to the readers.

Next, we would like to strengthen the above. To this end, we present the following lemma.

Lemma 5.2. Let $a_1, \cdots, a_n \in \mathbb{R}^m$ and $K(a_1, \cdots, a_n)$ be the convex cone generated by $a_1, \cdots, a_n$, i.e.,

$$K(a_1, \cdots, a_n) = \Big\{ \sum_{i=1}^n x_i a_i \mid x_i \geqslant 0, \quad 1 \leqslant i \leqslant n \Big\}.$$

Then $K(a_1, \cdots, a_n)$ is closed.

Proof. Let $A = (a_1, \cdots, a_n) \in \mathbb{R}^{m \times n}$, and recall $\mathbb{R}_+^n \triangleq \{ x \in \mathbb{R}^n \mid x \geqslant 0 \}$. Then

$$K(a_1, \cdots, a_n) = A(\mathbb{R}_+^n).$$

Now, suppose $y_k \in A(\mathbb{R}_+^n)$ with $y_k \to \bar{y}$ for some $\bar{y} \in \mathbb{R}^m$. We need to show that $\bar{y} \in A(\mathbb{R}_+^n)$. Let $x_k \in \mathbb{R}_+^n$ such that

$$y_k = A x_k, \qquad k \geqslant 1.$$

Suppose

$$\text{rank}(A) = \ell \leqslant m.$$

Then there exists an invertible matrix $P \in \mathbb{R}^{m \times m}$ such that

$$PA = \begin{pmatrix} \bar{A} \\ 0 \end{pmatrix}, \qquad \bar{A} \in \mathbb{R}^{\ell \times n}, \qquad \text{rank}(\bar{A}) = \ell.$$

Consequently,

$$P y_k = P A x_k = \begin{pmatrix} \bar{A} x_k \\ 0 \end{pmatrix} \equiv \begin{pmatrix} z_k \\ 0 \end{pmatrix}, \qquad z_k \in \mathbb{R}^\ell.$$

Note that $z_k \geqslant 0$ might not be true. But, we may let $Q_k \in \mathbb{R}^{\ell \times \ell}$ be a diagonal matrix whose diagonal entries are ± 1 so that

$$Q_k z_k = \bar{z}_k \geqslant 0.$$

Since $\{y_k\}_{k \geqslant 1}$ is bounded, so is $\{\bar{z}_k\}_{k \geqslant 1}$ (noting that P is invertible). Now, denote $\mathbf{1} = (1, \cdots, 1)^\top \in \mathbb{R}^n$. For each $k \geqslant 1$, consider the following LP problem:

$$\begin{cases} \min & \mathbf{1}^\top x, \\ \text{subject to} & (Q_k \bar{A}) x = \bar{z}_k, \qquad x \geqslant 0. \end{cases} \tag{5.10}$$

Since the above LP problem is feasible and bounded below, we have an optimal solution $\bar{x}_k$. Thus, by Proposition 3.2, under a suitable basic decomposition, we have the representation

$$\bar{x}_k = \begin{pmatrix} \bar{B}_k^{-1} \bar{z}_k \\ 0 \end{pmatrix}.$$

Here, $\bar{B}_k$ is formed by suitably selecting ℓ columns of $Q_k \bar{A}$. Thus it is depending on z_k and

$$\bar{B}_k^{-1} = (Q_k B_k)^{-1} = B_k^{-1} Q_k^{-1} = B_k^{-1} Q_k,$$

for some

$$B_k \in \mathcal{B} \equiv \{B \mid B \text{ is an invertible } (\ell \times \ell) \text{ submatrix of } \bar{A}\}.$$

Since $\mathcal{B}$ is a finite set and there are finitely many different Q_k $(k \geq 1)$, by choosing a subsequence if necessary, we may assume that

$$B_k = B, \quad Q_k = Q_1, \qquad \forall k \geq 1,$$

for some $B \in \mathcal{B}$. Then (noting $Q_1^{-1} = Q_1$)

$$\bar{x}_k = \begin{pmatrix} B^{-1} Q_1 \bar{z}_k \\ 0 \end{pmatrix} = \begin{pmatrix} B^{-1} \bar{z}_k \\ 0 \end{pmatrix}, \qquad k \geq 1.$$

Consequently, $\{\bar{x}_k\}_{k \geq 1}$ is a bounded subset in $\mathbb{R}_+^n$, and we may assume that $\bar{x}_k \to \bar{x} \in \mathbb{R}_+^n$ and $z_k \to \bar{z}$. Then

$$\bar{A}\bar{x} = \bar{z}, \qquad \bar{y} = A\bar{x} \in A(\mathbb{R}_+^n),$$

proving our conclusion. $\qquad\qquad\qquad\qquad\qquad\qquad\qquad\qquad\qquad\qquad$ □

The following is a very nice application of the above lemma and the separation theorem for convex sets.

Theorem 5.3. (Farkas) (i) *System*

$$Ax = b, \qquad x \geq 0 \tag{5.11}$$

has no solution if and only if system

$$A^\top w \leq 0, \qquad b^\top w > 0 \tag{5.12}$$

admits a solution.

(ii) *System* (5.11) *has no solution if and only if system*

$$A^\top w \geq 0, \qquad b^\top w = -1 \tag{5.13}$$

admits a solution.

(iii) *System*

$$Ax \geqslant b, \quad x \geqslant 0 \tag{5.14}$$

has no solution if and only if the system

$$A^\top w \leqslant 0, \quad b^\top w > 0, \quad w \geqslant 0 \tag{5.15}$$

admits a solution.

(iv) *System*

$$Ax \leqslant b, \quad x \geqslant 0 \tag{5.16}$$

has no solution if and only if the system

$$A^\top w \geqslant 0, \quad b^\top w = -1, \quad w \geqslant 0 \tag{5.17}$$

admits a solution.

Proof. (i) Let $A = (a_1, a_2, \cdots, a_n)$ with $a_i \in \mathbb{R}^m$ $(1 \leqslant i \leqslant n)$. Thus, (5.11) does not have a solution if and only if

$$b \neq \sum_{i=1}^{n} x_i a_i, \qquad \forall x_i \geqslant 0. \tag{5.18}$$

Namely, if we let $K(a_1, \cdots, a_n)$ be the convex cone generated by $a_1, \cdots, a_n$, then by Lemma 5.2, it is closed, and (5.18) is equivalent to the following:

$$b \notin K(a_1, \cdots, a_n).$$

Hence, by Theorem 1.5 of Chapter 4 (separation theorem for convex sets), we can find a $w \neq 0$ such that

$$w^\top b > 0, \qquad w^\top a_i \leqslant 0, \qquad 1 \leqslant i \leqslant n.$$

This proves (i).

(ii) We need only note that if (5.12) admits a solution w, then

$$\bar{w} \triangleq -\frac{w}{b^\top w}$$

is a solution of (5.13). Conversely, if (5.13) admits a solution $\bar{w}$, then $-\bar{w}$ is a solution of (5.11).

(iii) Introduce slack variable $y \geqslant 0$ to the system (5.14), namely, we consider system

$$Ax - y = (A, -I) \begin{pmatrix} x \\ y \end{pmatrix} = b, \quad \begin{pmatrix} x \\ y \end{pmatrix} \geqslant 0. \tag{5.19}$$

Then (5.14) admits no solution if and only if (5.19) admits no solution. Hence, by (i), it is equivalent to that the system

$$\begin{pmatrix} A^\top \\ -I \end{pmatrix} w \leqslant 0, \quad b^\top w > 0 \tag{5.20}$$

admits a solution. Clearly, (5.20) is the same as (5.15).

(iv) Introduce slack variable $y \geqslant 0$ to the system (5.16), namely, we consider system

$$Ax + y = (A, I) \begin{pmatrix} x \\ y \end{pmatrix} = b, \quad \begin{pmatrix} x \\ y \end{pmatrix} \geqslant 0. \tag{5.21}$$

Then (5.16) admits no solution if and only if (5.21) admits no solution. Hence, by (ii), it is equivalent to that the system

$$\begin{pmatrix} A^\top \\ I \end{pmatrix} w \geqslant 0, \quad b^\top w = -1 \tag{5.22}$$

admits a solution. Clearly, (5.22) is the same as (5.17). □

Parts (i)–(ii) are concerned with standard form LP, and Parts (iii)–(iv) are concerned with canonical form LP.

Now, we are ready to state and prove the following result which is a strengthened version of the weak duality principle.

Theorem 5.4. (Strong Duality Principle) *Consider standard form LP problem (1.1) and its dual (5.1).*

(i) If either (1.1) or (5.1) admits an optimal solution, so does the other and they have the same optimal objective value.

(ii) If either (1.1) or (5.1) is infeasible, then the other is either infeasible or unbounded.

Proof. (i) Suppose $x^* \in F_P$ is an optimal solution of primal LP problem (1.1) with optimal objective function value $z^* = c^\top x^*$. Define

$$K \triangleq \left\{ \begin{pmatrix} \rho \\ u \end{pmatrix} \in \mathbb{R}^{1+m} \;\middle|\; \begin{pmatrix} \rho \\ u \end{pmatrix} = \begin{pmatrix} z^* & -c^\top \\ b & -A \end{pmatrix} \begin{pmatrix} t \\ x \end{pmatrix}, \; \begin{pmatrix} t \\ x \end{pmatrix} \geqslant 0 \right\}.$$

Clearly, K is a convex cone. By Lemma 5.2, K is closed. Now, we claim that $(1, 0) \notin K$. In fact, if otherwise, then we have some $t_0 \geqslant 0$ and $x_0 \geqslant 0$ such that

$$t_0 z^* - c^\top x_0 = 1, \quad t_0 b - A x_0 = 0 \in \mathbb{R}^m. \tag{5.23}$$

If $t_0 = 0$, the above reads

$$Ax_0 = 0, \qquad c^\top x_0 = -1 < 0, \qquad x_0 \geqslant 0.$$

This implies that $x^* + \lambda x_0 \in F_P$ for any $\lambda > 0$. Hence,

$$c^\top(x^* + \lambda x_0) = z^* - \lambda \to -\infty, \qquad \lambda \to \infty,$$

contradicting the optimality of x^*. Thus, we must have $t_0 > 0$. Now, from the second relation in (5.23), we see that $\frac{x_0}{t_0} \in F_P$. Consequently, by the optimality of x^*,

$$1 = t_0 z^* - c^\top x_0 = t_0 \left(c^\top x^* - c^\top \frac{x_0}{t_0} \right) \leqslant 0,$$

a contradiction again. Therefore, $(1, 0) \notin K$.

Now, applying separation theorem for convex sets (Theorem 1.5 of Chapter 4), we can find a non-zero vector $(s, w) \in \mathbb{R}^{1+n}$ such that for some $\beta \in \mathbb{R}$,

$$s = s \cdot 1 + w^\top 0 < \beta = \inf\{s\rho + w^\top u \mid (\rho, u) \in K\}.$$

Since K is a cone, it is necessary that $\beta \geqslant 0$. Otherwise, there would be some $(\rho, u) \in K$ such that

$$s\rho + w^\top u < 0.$$

Then due to the fact that $\lambda(\rho, u) \in K$ for all $\lambda > 0$, we obtain

$$s < \beta \leqslant \lambda(s\rho + w^\top u) \to -\infty, \qquad \lambda \to \infty,$$

a contradiction. On the other hand, since $(0, 0) \in K$, one has $\beta \leqslant 0$. Hence, we must have $\beta = 0$. Then $s < 0$. Without loss of generality, we may assume that $s = -1$. Thus, we obtain some $w \in \mathbb{R}^n$ such that

$$-\rho + w^\top u \geqslant \beta = 0, \qquad \forall(\rho, u) \in K,$$

or, equivalently, by the definition of K,

$$\rho = tz^* - c^\top x, \qquad u = tb - Ax,$$

we have

$$(c - A^\top w)^\top x + t(b^\top w - z^*) \geqslant 0, \qquad \forall t \geqslant 0, \ x \geqslant 0.$$

Take $t = 0$, we obtain

$$A^\top w \leqslant c.$$

Thus $w \in F_D$, leading to (making use of the weak duality principle)

$$c^\top x^* \geqslant b^\top w.$$

Then set $x = 0$ and $t = 1$, one has (noting the definition of z^*)

$$b^\top w \geqslant z^* = c^\top x^* \geqslant b^\top w.$$

Hence, the equalities in the above hold. Then by Theorem 5.1, w is optimal for the dual LP problem.

Conversely, suppose $w^* \in F_D$ is an optimal solution of dual LP (5.1). We define

$$\begin{cases} u^* \equiv (u_1^*, \cdots, u_m^*)^\top, & u_i^* = \max\{w_i^*, 0\}, \\ v^* \equiv (v_1^*, \cdots, v_m^*)^\top, & v_i^* = \max\{-w_i^*, 0\}, \\ s^* = c - A^\top w^*. \end{cases} \tag{5.24}$$

Then $w^* = u^* - v^*$ and $y^* \equiv (u^*, v^*, s^*)^\top$ is an optimal solution of (5.3), which is in its standard form. Hence, by what we have proved, the dual of (5.3), which is equivalent to the primal LP problem (1.1), admits an optimal solution with the same optimal objective function value. This proves (i).

(ii) Suppose the primal problem is infeasible. If the dual problem is infeasible as well, we are done. Otherwise, let $w \in F_D$. By Theorem 5.3 (i), since $Ax = b$, $x \geqslant 0$ has no solution, there exists a $\bar{w}$ such that

$$A^\top \bar{w} \leqslant 0, \qquad b^\top \bar{w} > 0. \tag{5.25}$$

Then for all $\lambda \geqslant 0$, $w + \lambda \bar{w} \in F_D$ and clearly, $b^\top(w + \lambda \bar{w})$ is unbounded above. This means that the dual problem is unbounded. Next, let the dual problem be infeasible, then by what we have proved, its dual (which is the primal problem) will be either infeasible or unbounded. $\qquad \square$

In the case that x^* is non-degenerate, we may prove Theorem 5.4 (i) in the following way (without involving separation theorem for convex sets). Assume that x^* is given by the following:

$$x^* = \begin{pmatrix} B^{-1}b \\ 0 \end{pmatrix}, \qquad B^{-1}b > 0,$$

where $A = (B, N)$, and correspondingly $c = \begin{pmatrix} c_B \\ c_N \end{pmatrix}$. Let

$$w^* = [B^{-1}]^\top c_B, \tag{5.26}$$

which is the shadow price of the problem, also called a *simplex multiplier* corresponding to the given basic decomposition. Then (noting Proposition 3.2, and the non-degeneracy of x^*)

$$c - A^\top w^* = \begin{pmatrix} c_B \\ c_N \end{pmatrix} - \begin{pmatrix} B^\top \\ N^\top \end{pmatrix} w^* = \begin{pmatrix} 0 \\ c_N - (B^{-1}N)^\top c_B \end{pmatrix} \equiv r \geqslant 0.$$

This means that $w^* \in F_D$. Moreover,

$$c^\top x^* = c_B^\top x_B^* = c_B^\top B^{-1} b = (w^*)^\top b = b^\top w^*.$$

Hence, by Theorem 5.1, w^* is an optimal solution of dual LP problem (5.1). Note, however, that when x^* is degenerate, we do not know if the simplex multiplier w^* defined by (5.26) is in F_D. Thus, the above argument does not go through. But the above idea will be very useful, and we will return to this shortly.

The following is referred to as the *complementary slackness theorem*.

Theorem 5.5. *Consider canonical form LP problem (1.5) and its dual LP problem (5.5). Suppose $x \in F_P$ and $w \in F_D$, respectively. Then x and w are respectively optimal solutions of (1.5) and (5.5) if and only if*

$$\begin{cases} (c - A^\top w)^\top x = 0, \\ (Ax - b)^\top w = 0. \end{cases} \tag{5.27}$$

Proof. Define the *primal slackness vector*

$$s = Ax - b \geqslant 0,$$

and the *dual slackness vector*

$$\rho = c - A^\top w \geqslant 0.$$

Now, for any $x \in F_P$ and $w \in F_D$, we have

$$0 \leqslant \rho^\top x + s^\top w = (c - A^\top w)^\top x + w^\top (Ax - b) = c^\top x - b^\top w. \tag{5.28}$$

Hence, x and w are respectively optimal to primal and dual LP problem (1.5) and (5.5), which is equivalent to $c^\top x = b^\top w$, if and only if

$$\rho^\top x = 0, \qquad s^\top w = 0, \tag{5.29}$$

proving our claim. □

Conditions (5.29) are referred to as *complementary slackness*. Now, for standard form LP problem (1.1), we have the following *KKT condition*, which is a corollary of the above.

Theorem 5.6. *An x is optimal for standard LP problem (1.1) if and only if there exists a w such that*

$$\begin{cases} Ax = b, \qquad x \geqslant 0, & \text{(primal feasibility)}, \\ c - A^\top w \geqslant 0, & \text{(dual feasibility)}, \\ (c - A^\top w)^\top x = 0, & \text{(complementary slackness)}. \end{cases} \tag{5.30}$$

Proof. Suppose x is an optimal solution to LP problem (1.1). Let

$$L(x, \lambda, \mu) = c^\top x + \lambda^\top (Ax - b) - \mu^\top x.$$

Then by KKT condition, one should have

$$\begin{cases} 0 = L_x(x, \lambda, \mu) = c^\top + \lambda^\top A - \mu^\top, \\ \mu \geqslant 0, \qquad \mu^\top x = 0. \end{cases}$$

Let $w = -\lambda$. Then we have

$$c - A^\top w = c + A^\top \lambda = \mu \geqslant 0,$$

and

$$(c - A^\top w)^\top x = \mu^\top x = 0.$$

This proves (5.30).

Conversely, let $x, w \in \mathbb{R}^n$ satisfy (5.30). Then $w \in F_D$ and

$$c^\top x - b^\top w = c^\top x - w^\top (Ax) = (c - A^\top w)^\top x = 0.$$

Hence, by Theorem 5.1, we see that x is an optimal solution of LP problem (1.1). □

Now, for a pair of primal and dual problems, we define

$$z_P^* = \inf_{x \in F_P} c^\top x, \qquad z_D^* = \sup_{w \in F_D} b^\top w. \tag{5.31}$$

As a convention, we let

$$\inf \varnothing = +\infty, \qquad \sup \varnothing = -\infty.$$

Similar to Definition 3.2 of Chapter 4, we say that the primal-dual pair problems have no gap if

$$z_P^* = z_D^*,$$

and say that the primal-dual pair problems have a gap if

$$z_P^* > z_D^*.$$

The following examples are relevant to the duality gaps.

Example 5.7. Consider

$$\begin{cases} \min & -2x_1 + x_2, \\ \text{subject to} & \begin{cases} x_1 - x_2 + x_3 = 1, \\ x_1 - x_2 - x_4 = 2, \\ x_1, x_2, x_3, x_4 \geqslant 0. \end{cases} \end{cases}$$

From the above, we see that

$$1 = x_1 - x_2 + x_3 \geqslant x_1 - x_2 \geqslant x_1 - x_2 - x_4 = 2,$$

which is a contradiction. Thus, $F_P = \varnothing$. Note that

$$A = \begin{pmatrix} 1 & -1 & 1 & 0 \\ 1 & -1 & 0 & -1 \end{pmatrix}, \quad b = \begin{pmatrix} 1 \\ 2 \end{pmatrix}, \quad c^\top = (-2, 1, 0, 0).$$

Therefore, the dual problem reads

$$\begin{cases} \max \quad w_1 + 2w_2, \\ \\ \text{subject to} \begin{cases} w_1 + w_2 \leqslant -2, \\ -w_1 - w_2 \leqslant 1, \\ w_1 \leqslant 0, \\ -w_2 \leqslant 0. \end{cases} \end{cases}$$

Then

$$-1 \leqslant w_1 + w_2 \leqslant -2,$$

which is a contradiction. Hence, $F_D = \varnothing$. Consequently, by our convention,

$$z_P^* = \inf_{x \in F_P} c^\top x = +\infty, \qquad z_D^* = \sup_{w \in F_D} b^\top w = -\infty.$$

Thus the gap is infinite.

Example 5.8. Consider

$$\begin{cases} \min \quad x_1 - 2x_2, \\ \\ \text{subject to} \begin{cases} x_1 - x_2 + x_3 = 1, \\ -x_1 + 2x_2 - x_4 = 2, \\ x_1, x_2, x_3, x_4 \geqslant 0. \end{cases} \end{cases}$$

For this problem, we have

$$F_P = \left\{ \begin{pmatrix} x_2 - x_3 + 1 \\ x_2 \\ x_3 \\ x_2 + x_3 - 3 \end{pmatrix} \;\middle|\; x_2, x_3 \geqslant 0, \; x_2 - x_3 + 1 \geqslant 0, \; x_2 + x_3 - 3 \geqslant 0 \right\}.$$

Clearly,

$$x = \begin{pmatrix} 0 \\ x_2 \\ x_2 + 1 \\ 2(x_2 - 1) \end{pmatrix} \in F_P, \qquad \forall x_2 \geqslant 1,$$

and for such a solution,

$$c^\top x = -2x_2.$$

Hence, the problem is unbounded. Then by Theorem 5.1, we must have that the dual problem is infeasible. In fact, in the current case,

$$A = \begin{pmatrix} 1 & -1 & 1 & 0 \\ -1 & 2 & 0 & 1 \end{pmatrix}, \quad b = \begin{pmatrix} 1 \\ 2 \end{pmatrix}, \quad c^\top = (1, -2, 0, 0).$$

Thus, the dual problem is

$$\begin{cases} \max & w_1 + 2w_2, \\ \\ \text{subject to} & \begin{cases} w_1 - w_2 \leqslant 1, \\ -w_1 + w_2 \leqslant -2, \\ w_1 \leqslant 0, \\ w_2 \leqslant 0. \end{cases} \end{cases}$$

For any $w \in F_D$,

$$2 \leqslant w_1 - w_2 \leqslant 1,$$

which is a contradiction. Hence, $F_D = \varnothing$. By our convention,

$$z_P^* = \inf_{x \in F_P} c^\top x = -\infty, \qquad z_D^* = \sup_{w \in F_D} b^\top w = -\infty.$$

For this case, the primal-dual pair of problems have no gap.

Example 5.9. Consider

$$\begin{cases} \min & x_1 + 2x_2 + 5x_3, \\ \\ \text{subject to} & \begin{cases} -x_1 - x_2 + x_3 = 1, \\ -x_1 - 2x_2 - x_3 = 1, \\ x_1, x_2, x_3 \geqslant 0. \end{cases} \end{cases}$$

If there exists an $x \in F_P$, then

$$0 \geqslant -2x_1 - 3x_2 = 2,$$

which is a contradiction. Thus, $F_D = \varnothing$, and

$$z_P^* = \inf_{x \in F_D} c^\top x = +\infty.$$

Now,

$$A = \begin{pmatrix} -1 & -1 & 1 \\ -1 & -2 & -1 \end{pmatrix}, \quad b = \begin{pmatrix} 1 \\ 1 \end{pmatrix}, \quad c^\top = (1, 2, 5).$$

Thus, the dual problem is

$$\begin{cases} \max \quad w_1 + w_2, \\ \text{subject to} \quad \begin{cases} -w_1 - w_2 \leqslant 1, \\ -w_1 - 2w_2 \leqslant 2, \\ w_1 - w_2 \leqslant 5. \end{cases} \end{cases}$$

Clearly, $w = \lambda(1,1) \in F_D$ for any $\lambda \geqslant 0$. Hence,

$$z_D^* = \sup_{w \in F_D} b^\top w = +\infty.$$

Consequently, the primal-dual pair problems have no gap.

The above three examples exhibit that the following cases are possible:

$$\begin{aligned} F_D = \varnothing, && +\infty = z_P^* > z_D^* = -\infty, \\ F_D = \varnothing, && z_P^* = z_D^* = -\infty, \\ F_D \neq \varnothing, && z_P^* = z_D^* = \infty. \end{aligned}$$

Exercises

1. For primal problem of canonical form (1.5) and its dual problem (5.5), state and prove the weak duality principal and strong duality principal.

2. Consider the following LP problem

$$\begin{cases} \min \quad 2x_1 - 3x_2 + x_3, \\ \text{subject to} \quad \begin{cases} 2x_1 + 4x_2 - 3x_3 = 1, \\ -3x_1 - 2x_3 + x_4 = 3, \\ x_1, x_2, x_3, x_4 \geqslant 0. \end{cases} \end{cases}$$

Find the dual problem. Is there a gap between the primal and dual problems?

Bibliography

[1] L. D. Berkovitz, *Convexity and Optimization in* $\mathbb{R}^n$, John Wiley & Sons, Inc., New York, 2002.

[2] E. K. P. Chong and S. H. Zak, *An Introduction to Optimization*, 4th Ed., John Wiley & Sons, Inc., Hoboken, 2013.

[3] G. B. Dantzig, *Maximization of a linear function of variables subject to linear inequalities, Activity Analysis of Prduction and Allocation*, 339–347. Cowles Commission Monograph No.13, John Wiley & Sons, New York, 1951.

[4] I. Ekeland, *On the variational principle*, J. Math. Anal. Appl., 47 (1974), 324–353.

[5] S.-P. Han and O. L. Mangasarian, *Exact penalty functions in nonlinear programming*, Math. Programming, 17 (1979), 251–269.

[6] F. John, *Extremum problems with inequalities as side conditions*, Studies and Essays, Courant Anniversary Volume, K. O. Friedrichs, O. E. Neugebauer and J. J. Stoker, eds, Wiley, New York, 1948, pp.187–204.

[7] W. Karush, *Minima of Functions of Several Variables with Inequalities as Side Conditions*, Masters Thesis, University of Chicago, 1939.

[8] H. W. Kuhn and A. W. Tucker, *Nonlinear programming*, Proceedings of the 2nd Berkley Symposium, University of California Press, 1951, 481–492.

[9] D. G. Luenberger and Y. Ye, *Linear and Nonlinear Programming*, 3rd Ed., Springer, 2008.

[10] O. L. Mangasarian and S. Fromovitz, *The Fritz John necessary optimality conditions in the presence of equality and inequality constraints*, J. Math. Anal. Appl., 17 (1967), 37–47.

[11] G. P. McCormick, *Second order conditions for constrained minima*, SIAM J. Appl. Math., 15 (1967), 641–652.

[12] R. K. Sundaram, *A First Course in Optimization Theory*, Cambridge Univ. Press, 1996.

Index

Printed in the United States
By Bookmasters